The Ethical Chemist

Professionalism and Ethics in Science

The Ethical Chemist

Professionalism and Ethics in Science

Jeffrey Kovac

University of Tennessee

Prentice Hall Series in Educational Innovation

Pearson Education, Inc,
Upper Saddle River, NJ 07458

CIP data on file

Senior Acquisitions Editor: *Kent Porter-Hamann*
Editorial Assistant: *Jacquelyn M. Howard*
Vice President and Director of Production and Manufacturing, ESM: *David W. Riccardi*
Production Editor: *Colleen Franciscus*
Director of Creative Services: *Paul Belfanti*
Creative Director: *Carole Anson*
Art Director: *Jayne Conte*
Cover Designer: *Bruce Kenselaar*
Manufacturing Manager: *Trudy Pisciotti*
Manufacturing Buyer: *Lynda Castillo*
Marketing Manager: *Steve Sartori*

PEARSON
Prentice
Hall

©2004 by Pearson Education, Inc.
Pearson Education, Inc.
Upper Saddle River, NJ 07458

Printed in the United States of America

10 9 8 7 6 5 4 3

ISBN 0-13-141132-2

Pearson Education Ltd., *London*
Pearson Education Australia Pty. Ltd., *Sydney*
Pearson Education Singapore, Pte. Ltd.
Pearson Education North Asia Ltd., *Hong Kong*
Pearson Education Canada, Inc., *Toronto*
Pearson Education de Mexico, S.A. de C.V.
Pearson Education—Japan, *Tokyo*
Pearson Education Malaysia, Pte. Ltd.
Pearson Education Inc., *Upper Saddle River, New Jersey*

To Peter and Rachel
and to the memory of Charles Davis

Contents

Preface

As a child growing up during the coldest part of the Cold War and fascinated by science, I began to wonder about the scientists who created the terrible nuclear weapons that threatened to annihilate humanity. What did they think about the consequences of their work? In retrospect, this was the beginning of my interest in scientific ethics. In high school I discovered Jacob Bronowski's wonderful little book, *Science and Human Values* which helped me begin thinking about the question more systematically. Although I was a chemistry major in college, I took several philosophy courses, including ethics, and continued to read informally about the ethics of science. This was the era of the Vietnam War, so ethical questions such as the morality of using herbicides, the infamous "Agent Orange," and napalm were part of the public discourse. There was much to consider.

As a graduate student I learned how to be a professional scientist, but the larger context of science was never far from my mind. At Yale auditing Martin Klein's courses in the history of science furthered my interest. As a young faculty member I focused on building a scientific career, but was finally able to put my interests in history and philosophy of science to use professionally at the University of Tennessee in 1988 when I taught a capstone course for senior chemistry majors that was supposed to explore the historical and cultural context of chemistry. In developing this course over seven years I began to introduce questions of scientific integrity. Ethical issues were in the news at the time, so the daily press and *Science* and *Chemical and Engineering News* provided plenty of material. Although the literature on biomedical ethics was emerging, there was little to be found on ethical questions in physical science. To help fill that gap, in 1993 (rev. 1995) I wrote a casebook, *The Ethical Chemist*, with the support of the Camille and Henry Dreyfus Foundation. This casebook has been available for purchase at cost from the Department of Chemistry at the University of Tennessee. I have lost track of how many copies we have distributed, but it is certainly more than five hundred.

Since 1995 I have been exploring questions of scientific and professional ethics in more detail, developing my thinking in a series of articles that have been published in a number of venues. In addition, several users of the casebook have suggested that it be revised and expanded for a larger audience. The original version was written for an audience of senior chemistry majors and beginning graduate students, but it has been used, with varying success, in other contexts including lower division undergraduate courses and corporate settings. In rereading the original casebook I realized that many things could have been done much better. Writing the book you have in hand has provided an opportunity to develop my ideas about professionalism and ethics in science more fully, to revise the cases and commentaries in *The Ethical Chemist*, and to add a number of new cases.

I have tried to write a self-contained introduction to professional ethics for both chemistry students and practicing chemists. It can also be used as a textbook for a course or seminar in scientific ethics and as an instructor resource. The individual cases can be used as prompts for class discussions or writing assignments in many of the usual courses in an undergraduate or graduate chemistry curriculum. The Committee on Professional Training of the American Chemical Society has recently recommended that education in professional ethics be included in the undergraduate chemistry curriculum and has published suggestions for implementing this recommendation. This book provides adequate material for any of these possibilities. While it is written for chemists, the cases easily can be adapted for other sciences.

Many people have helped with this project. Over the years, I have been fortunate to have had a number of talented undergraduate research students working on it with me. The original version of *The Ethical Chemist* would not have been completed without the enthusiasm and hard work of Priscilla A. Frase. Kristy Carter wrote preliminary versions of several cases, and Sean Seymore and Schylon Yates did important background research for that book. The present volume has benefitted from the excellent work of Michael Bleakley, Melinda Coker, Rachel Graves, Jennifer J. Rosenbaum, and especially Jason Johnson. Anne Moody of Truman State University contributed drafts of several new cases appropriate for use in lower-division chemistry courses. At various times The Ronald McNair Postbaccalaureate Achievement Program, the University of Tennessee, and especially the Camille and Henry Dreyfus Foundation have provided financial support. Special thanks to Robert L. Lichter, the former director of the Dreyfus Foundation, for his personal support of my work.

In pursuing an interdisciplinary project it is important to talk to and to learn from colleagues. My mentors in chemistry, William D. Weir, Marshall Fixman, Irwin Oppenheim, and John W. Larsen, not only helped me learn how to do high-quality research but also were examples of scientific integrity. My undergraduate education at Reed College provided the broad perspective on the liberal arts necessary for undertaking a project like this. Over the past ten years I have benefitted enormously from discussions and correspondence with Davis Baird, Linda Bensel-Meyers, Norman S. Care, Brian P. Coppola, Michael Davis, Donald Gotterbarn, Roald Hoffmann, and Linda Sweeting. I am particularly grateful to Roger Jones from whom I have received both minor suggestions and major enlightenment during our twenty-five-year dialogue about philosophy and science. Donna W. Sherwood, my friend and colleague was a superb copy editor. Kent Porter-Hamann, Senior Editor, John Challice, Editor-in-Chief, Jacquelyn Howard, Editorial Assistant, and Lynda Castillo, at Prentice Hall made this book a reality. It is impossible adequately to thank my wife, Susan Davis Kovac, for her intellectual contributions and patient and loving support. Finally, this book is dedicated to my children, Peter and Rachel, and to the memory of Charles Davis, three remarkable examples of integrity and moral courage.

Jeffrey Kovac

About the Author

Educated at Reed College and Yale University, **Jeffrey Kovac** is Professor of Chemistry at the University of Tennessee where he has taught undergraduate and graduate courses in chemistry and interdisciplinary courses for the university honors program since 1976. Since 1994 he has served as Director of the Tennessee Governor's School for the Sciences, a four-week summer residential enrichment program in science for talented high school students. He is a Fellow of the American Association for the Advancement of Science. His interdisciplinary scholarly interests include statistical mechanics and thermodynamics, history and philosophy of science, especially scientific ethics, and the scholarship of teaching and learning. He has been writing and lecturing about ethics in science for more than ten years. He is coauthor, along with Donna W. Sherwood, of *Writing Across the Chemistry Curriculum: An Instructor's Handbook*, also published by Prentice Hall (2001).

The Ethical Chemist

Professionalism and Ethics in Science

Chapter 1

Introduction

The chemist is "both a craftsman and a philosopher" (Knight 1992, 13). Chemistry traces its origins to ancient crafts such as metalworking, dyeing, tanning, and ceramics and also to the ancient philosophers' speculations about the nature of matter. The practice of chemistry is a unique combination of the theoretical and the practical that raises interesting ethical questions. Chemistry has transformed the modern world with amazing new materials, powerful drugs, agricultural products that have increased farm productivity, new and better explosives that can be used for both construction and destruction, synthetic textile fibers, brilliantly colored dyes, and countless others. With these advances have come unfortunate effects such as environmental pollution. As a result, moral questions concerning the relationship between chemistry and society have become increasingly urgent (Hoffmann 1997).

The remarkable scientific successes during World War II, such as the Manhattan Project and the development of radar, changed the practice of science. Science, including chemistry, was no longer a leisurely activity. Government funding, epitomized by the founding of the National Science Foundation, along with increased private foundation and industry support stimulated the growth of research programs in universities, national laboratories, and private research institutions. Scientific research has become high profile and high pressure; the rewards for success can be significant, in both prestige and money. While scientists have always been competitive, the culture of the community has changed in recent years, straining the bonds of collegiality and bringing questions of professional ethics to the forefront.

In 1933 the distinguished physicist Freeman J. Dyson suggested that science was "in trouble" at three levels: personal, local, and global (Dyson 1993). At the core of Dyson's analysis were questions of ethics, both the personal integrity of individual scientists and the relationship between science and society. Around the same time, the National Academy of Sciences issued *Responsible Science: Ensuring the Integrity of the Research Process*, a two-volume comprehensive study of scientific integrity (Panel on Scientific Responsibility and the Conduct of Research 1992, 1993). The appearance of this report highlighted the magnitude of concern about ethics within the scientific community. Responses to the concern have included conferences and symposia on ethics, the development of new courses and seminars on research ethics, summer seminars for both faculty and graduate students on the teaching of ethics, and new policies and procedures on research integrity and scientific misconduct at universities and national laboratories. Many excellent books and articles have been published. Much attention has been paid to the biological sciences, computer science, and engineering, but much less attention to the physical sciences, including chemistry. The present volume is an attempt to fill this void.

My view is that ethics is integral to science. From the personal to the global level, science is filled with moral decisions, large and small, so scientists must understand the ethical dimensions of situations and events and become sophisticated moral decision makers. Therefore this book begins with a discussion of ethics and ethical theory to provide a context for a thorough exposition of professionalism and ethics in chemistry. A general method for ethical problem solving is then presented and is followed by a series of real-life cases that raise the kinds of day-to-day ethical questions that working chemists confront. These cases are the "end-of-chapter problems," applying the general principles to specific situations. I have also included commentaries that offer "solutions."

Chapter 2 is a brief introduction to ethics and ethical theories and relies heavily on the concept of the common morality, the standards of conduct that are broadly shared in society. Several historically important ethical theories are discussed because each provides a different perspective on ethical reasoning. For example, consequentialist ethical theories, particularly utilitarianism, judge actions as good or bad based on their outcome. Deontological ethical theories, however, focus on the rightness or wrongness of the individual action regardless of the consequences. For a deontologist, "Don't lie" is a moral rule that should not be broken (although some exceptions might be permitted), but a consequentialist might allow lying in a circumstance in which the lie brought about more good than harm. In everyday life we use both kinds of reasoning to justify our actions, and both are useful ways to look at questions of professional ethics as well.

We usually think of ethics as universal, applying to all people, but professional ethics is specific; it applies only to a particular group of people. For example, legal ethics applies only to lawyers; scientific ethics applies only to scientists. Chapter 3 explores the concept of professional ethics in detail and lays out the sources and principles of the professional ethics of science in general and chemistry in particular. From the perspective that the standards of professional ethics develop historically based on a defining moral ideal, I try to elaborate the largely implicit standards of professional ethics that regulate the practice of chemistry.

Chapter 4 presents a four-step method for solving ethical problems. Chemists and students of chemistry may recognize many aspects of this method as analogous to the method for solving design problems, problems of making or repairing that satisfy human desires or needs. Chemistry is full of design problems; the design of a chemical synthesis is the most familiar. This methodology can be applied to the cases that make up Chapter 5.

The cases focus mainly on small issues, questions that arise in the routine practice of science. I have avoided the elaborate cases of scientific misconduct and fraud. Not only have major scandals received adequate coverage elsewhere, they also tend to be overwhelming. My experience in teaching professional ethics has reinforced my view that people learn ethical behavior best by making moral decisions, reflecting on the decision-making process, and thinking about why those decisions are right or wrong. As I note in Chapter 4, ethical problems are not like mathematics problems. Instead they are more like design problems for which the agent must devise and evaluate various possible courses of action. Just as there might be several good ways to make a particular compound, there also may be several morally acceptable courses of action. There is not always a single right thing to do. Some alternatives may also be clearly wrong because they violate a fundamental moral principle or rule or lead to unacceptable consequences.

Although this book is designed to be a stand-alone introduction to professional ethics in chemistry that can be read by an individual, it works best to discuss the cases with a group, either in a class or seminar, or more informally with a group of colleagues or friends. Groups are powerful problem-solving machines, and the different perspectives are useful in considering ethical problems. One of the most interesting features of ethical problems is that people see them quite differently depending on their personal experiences and values. Moreover, an important part of ethical decision making is giving good reasons for a decision. A discussion group can provide useful critique.

My goal has been to provide a broad introduction to ethics in chemistry ranging from the theoretical to the practical. I have tried to show that most, if not all, decisions in science have ethical dimensions. Thus, no one can be properly educated in chemistry or any other science without understanding the basis of professional ethics and learning the art of moral decision making.

Chapter 2

Ethics, Morals, and Ethical Theory

In ordinary language the words *ethics* and *morals* are used interchangeably to refer to standards of conduct or social norms that guide proper behavior. The English *ethics* derives from the Greek *ethika* meaning character or custom and is related to the Latin *mores*, also meaning custom, which gave us the word *moral*. Some philosophers, however, distinguish between the two. Morals is often taken to refer to universal norms of human behavior, the distinction between good and evil, while ethics is used as a generic term for all the different ways that scholars use to understand and examine our moral lives (Beauchamp and Childress 2001).

Some approaches to ethics are normative while others are non-normative. Normative approaches seek to discover and justify the general standards of behavior that we should accept and to apply them to specific situations. Non-normative approaches can be descriptive, that is, factual investigations of moral conduct and belief, or what is called meta-ethics, the analysis of ethical language, concepts, and methods of reasoning.

Morality generally refers to norms for right and wrong human conduct that are so widely shared that they form a stable social consensus. Here it is important to distinguish between what some philosophers call the common morality (the norms that all serious persons share) and communal norms that are only shared by a specific community. Common morality, although it cannot be precisely specified, is universal. Communal norms are similar to the common morality but are specific to a particular group, such as a religious or cultural community. Not part of the common morality, but still important, are moral ideals and extraordinary virtues, which call us to morally exemplary behavior.

Common Morality

Common morality seems to spring from human nature as shaped by living together in community. Successful communal life requires that people adhere to certain standards of behavior. For example, a principle of promise keeping seems essential to any society, whatever its specific organization. A principle of truth telling seems essential to all human relationships. If morality is universal, then it is important to try to learn why, and the numerous approaches to its origins include the study of its biological and social origins, which has recently become an active, though controversial, area of research (see, for example, Hinde 2002; Wright 1994). However, in this book we do not explore further the ultimate origins of morality.

We learn common morality both from our experiences and from our culture. An essential part of our moral experience is what William Gass (1980) calls *clear cases*, behavior that is clearly wrong or right. Some actions, such as torture and abuse of children, are obviously wrong. We would regard any person who condoned such behavior as morally deficient and would reject any moral principle that approved of such actions. Clear cases are the "experimental facts" of the common morality. From these clear cases come rules and principles, empirical generalizations that prescribe and proscribe our behavior. Finally, we have the intellectual heritage of a dialogue about ethics that goes back at least to ancient Greece. Society distills these different sources into the broadly accepted norms that comprise the common morality and that we learn as children from our parents, teachers, and religious leaders (Bok 1995).

Common morality can be summarized in terms of fundamental moral principles that express the general values underlying the more specific rules governing behavior. These moral principles include respect for autonomy, non-maleficence (avoidance of causing harm), beneficence (providing benefits), and justice (fair distribution of goods and services), among others (Beauchamp and Childress 2001). Moral rules are more specific guides to action, which can be expressed negatively ("Don't lie") or positively ("Tell the truth") (Gert 1988). In general, we think of moral rules as being substantive; that is, they express some important norm. But there are also procedural rules to help in making moral decisions when the substantive rules do not give a clear answer, and there are authority rules to specify who is entitled to make a decision. For example, common morality gives parents the authority to make certain decisions for their children.

Another way of thinking about morality, which goes back to Aristotle, uses the language of character and virtues. The various moral principles and rules have corresponding virtues. For example, the moral rule "don't lie" or "tell the truth" corresponds to the virtue of truthfulness. The principle of justice corresponds to the virtue of justice or perhaps fairness. Virtues refer to character traits that we hope individuals, and collectively societies, possess. Virtues lead to correct action. Beyond what might be called ordinary virtues are extraordinary virtues or supererogatory acts, which exceed what is normally expected of a morally responsible person and might be termed moral heroism. For example, dashing into a burning building to rescue a perfect stranger goes beyond what most people would consider a normal moral obligation, so a person who performs such an action is regarded as a (moral) hero.

Although common morality is something we all understand, it is not systematic. Philosophers have attempted to provide detailed accounts of the moral life and its context in what are called ethical theories (Beauchamp 1991; Rachels 1999). An ethical theory is similar to a scientific theory, although there are differences. A scientific theory is supposed to provide some sort of explanation of a class of natural phenomena. An ethical theory is a systematic presentation of the basic components of ethics derived from an integrated set of principles that is supposed to provide ultimate reasons for moral decisions. Both scientific and ethical theories are testable. A scientific theory must provide clear predictions that can be experimentally tested. Similarly, the experimental facts of ethics are the clear cases; any ethical theory that condones actions that are clearly wrong must either be rejected or modified. On the other hand, ethical theory is prescriptive in a way that scientific theory cannot be. Ethical theories provide both explanation and guidance for behavior. In Western culture, ethical theory goes back to Plato and Aristotle (MacIntyre 1966). Although there are many different variations, several major categories dominate ethical thinking. Although it is not essential to understand the various theories in detail, which is a formidable task, a basic understanding is useful because the theories represent the different ways that moral arguments are constructed. Each provides a different perspective corresponding to the different intuitions that we have about moral decisions. The theories provide different vocabularies and central concepts that appear over and over again in our day-to-day discussions about ethics as well as the different kinds of reasons that we use to plan and justify our actions.

One major school of thought is called consequentialism. Consequentialists evaluate actions as morally good or bad based on an assessment of their consequences. The most important consequentialist theory is utilitarianism, which is discussed later. Deontological theories are the second major class that is discussed. Deontological theories assert that certain binding principles or rules of conduct should be followed regardless of their consequences. While a consequentialist might decide whether lying in a certain circumstance is morally justified by asking whether the lie leads to a good result, a deontologist would argue that lying is *prima facie* immoral, so one should not lie even if the lie seems to produce significant benefits.

Two other kinds of theories have been historically influential: virtue theories focusing on the development of character; and communitarian theories taking a social contract approach. Virtue theories posit that morally desirable behavior comes from positive individual character traits called virtues rather than from following rules or evaluating consequences. In contrast, communitarian theories focus on the origins of morality in historical traditions and community decisions. Communitarian theories are important in understanding professional ethics, which derive from the values and practices of communities of professionals such as physicians and lawyers.

The following sections outline the four types of normative ethical theory: utilitarian, deontological, virtue, and communitarian. The philosophical ideas of the major proponents of each theory, along with the major objections to each, should provide sufficient background to help the reader understand how these different approaches can be used to analyze specific problems in personal and professional ethics.

Utilitarian Theories

Utilitarianism is a consequentialist ethical theory, meaning that actions are judged right or wrong according to their consequences, not to the motives of the actor. Using this theory, one ought to choose the action that would lead to the best consequences for all persons affected by it. Thus a certain action is better than a different one if it leads to the greatest possible balance of good consequences. It is important to note that utilitarianism is based on the implicit assumption that duty, obligation, and rights are less important than maximizing good or minimizing evil in a given context.

The two earliest significant utilitarian philosophers were Jeremy Bentham (1748–1832) and John Stuart Mill (1806–1873). Bentham was disenchanted with the foundations of contemporary British law, which viewed the law as the enforcement of divine commands. He argued that the purpose of law was to protect the welfare of citizens, and he saw utilitarianism as a practical system for legislators. Mill also proposed utilitarianism as a single standard for morals and legislation based on two foundations. First, Mill asserted the principle of utility as the normative foundation of his theory. He stated the principle in the following way: "Actions are right in proportion as they tend to promote happiness, wrong as they tend to produce the opposite of happiness" (Beauchamp 1991, 131). Second, he asserted a view of human nature that affirms a basic human desire for unity and harmony with fellow human beings. Following from these two foundations, the purpose of morality is seen as promoting natural human sympathies and combating human limitations. Mill and other utilitarians believe that the principle of utility is the best means to produce those ends.

Even within this school of thought there is disagreement on how to interpret the principle of utility. Bentham judged an action on its production of pleasure and non-production of pain.

He attempted to quantify this throughout the group affected by the action using criteria such as intensity, duration, propinquity, extent, and certainty. Mill tried to add a qualitative factor to decision making in response to a critique of utilitarianism now known as the "doctrine of swine objection": if only quantity is measured, a satisfied pig could become the moral standard. Mill refutes this idea in his famous comment that: "It is better to be a human being dissatisfied than a pig satisfied; better to be Socrates dissatisfied than a fool satisfied" (Beauchamp 1991, 134).

There are two different ways of applying the principle of utility: by act and by rule. *Act utilitarians* judge the consequences of individual actions independently, while *rule utilitarians* use the principle of utility to justify moral rules, which are then applied to individual actions. A rule utilitarian will obey a rule even in a situation where an act utilitarian would argue that breaking the rule would yield better consequences. Rule utilitarians argue that maintaining the integrity of the rules and the whole system of rules is important to the social fabric.

Philosophers have offered several strong critiques of utilitarianism. The first questions an individual's ability to weigh unlike things quantitatively or qualitatively as well as the individual's ability to weigh his or her own happiness equally with the happiness of others. Linked to this question is a pragmatic concern that even a person who had this ability would have insufficient time to calculate every decision. Mill has offered the rebuttal that conscience can be used as a rule of thumb to determine the rightness or wrongness of an action.

A second criticism argues that utilitarianism does not differentiate between obligatory acts, that is, acts that it would be morally wrong to omit, and supererogatory acts, which are morally good but are beyond the necessary moral obligation. Common morality includes such a distinction. Ordinary people are not condemned for not performing morally heroic acts such as going into a burning building to rescue strangers, but a utilitarian calculation might suggest that such actions are required. Utilitarians counter that because one's own happiness has to be taken into account in any calculation of utility, supererogatory acts will probably end up not being required. On the other hand, it is important to allow for supererogatory actions, so the principle of utility seems to break down as a general guide to action in this case.

A third criticism submits that utilitarianism requires one to subordinate private pursuits to the greater good. Utilitarians argue that the principle of utility only requires that one give up one's own pursuits if they are inconsistent with morality.

Utilitarian and other consequentialist theories are appealing because they correspond to our commonsense notion that the consequences of an action should make a difference in the way we behave. This is one of the ways we make decisions in our daily lives. The principle of utility, which calls for us to maximize the good and minimize the evil, is an essential part of the common morality. On the other hand, utilitarianism seems to lead to a situation where there are no absolutes, no rules, but the notion that some things are intrinsically good and bad is also part of our moral sense.

Deontological Theories

Deontology asserts that the rightness or wrongness of some actions is determined by binding patterns of conduct that are justified by more than consequences. Deontological theories vary in their dependence on consequences. Some declare that consequences are irrelevant to moral evaluations; others, barely distinguishable from the consequentialists, argue that only part of moral rightness is independent of consequential analysis. Perhaps the most widely known type of deontological theory is the *divine command theory*, in which the will of God is the ultimate standard of judgment. Other deontologists use reason to claim the moral value of actions. Additionally, intuition, or political, religious, or social contracts may be used.

Immanuel Kant, an eighteenth-century German philosopher, developed the most influential "monistic" deontological theory; monistic meaning that he envisioned one supreme principle, absolute and not supported by any other independent principle. Kant saw the ultimate basis for morality in reason. From this he asserted that all rational people possessed the same capacity for reason; thus moral rules apply universally. Asserting a pure form of deontology, Kant believed that to be moral an action must be performed for the sake of obligation. Kant asserted that no matter how intelligently one acts, the results of the actions are subject to accident and circumstance; thus, morality should be judged by its motivation and not its consequences.

For Kant, the supreme moral principle is a *categorical imperative* that persons ought to act only when they can will that the action become a universal moral law. He restates the principle: "One must act so as to treat every person as an end and never as a means only" (Beauchamp 1991, 188).

A strong critique of Kant's deontology questions how to resolve conflicting obligations. *Pluralistic* deontologists hold that there are several basic irreducible rules that provide "*prima facie* obligations." One is required to fulfill these obligations unless they are overridden by a stronger obligation. Turning Kant's absolute rules into *prima facie* rules leaves room to address conflicts and exceptions. Critics attack deontological pluralists on the grounds that they lack coherence and unity and that they fail to meet the basic goal of an ethical theory, to provide ultimate reasons for moral decisions. An additional critique of deontology asserts that it covertly appeals to the utilitarian principle.

Deontological theories also appeal to our moral common sense in suggesting that moral rules based on some absolute standard do exist. On the other hand, blind adherence to rules can lead to undesirable consequences. There are times when lying or breaking a promise seems to be appropriate, say in a situation where a life is at stake. One way to approach such situations is to balance principles, but then one needs some way to decide which principle is most important. One way to decide is by looking at consequences, but this is an unacceptable alternative for the pure deontologist.

Virtue Theories

Virtue ethics adopts the perspective that the goal of ethics is the cultivation of virtuous traits. Virtuous acts are defined not as moral requirements but as character traits that are socially and morally valued. Virtue ethics emphasizes motivation in judging an action; an individual must have both the disposition to act morally and the appropriate desire. If the action is taken only out of obligation and the desire is not right, then a necessary condition of virtue is lacking. In this regard, virtue theories differ from both utilitarian and deontological approaches.

Aristotle—as I have said before, one of the first ethical theorists—constructed a philosophical system constrained by a functional understanding of natural and social organization. He understood aspects of the world, including individuals, in terms of their natural functions and proper goals. Ancient Greek society was highly structured, so the idea of people's natural functions was for Aristotle a fairly straightforward concept. He explained virtue as "a disposition bred from an innate capacity by proper training and exercise of that capacity" (Beauchamp 1991, 217). Distinguishing between desire and reason, Aristotle divides virtue into *intellectual* and *practical* categories. He suggests that intellectual virtue may lead to vicious action, whereas an individual with the virtues of character and practical life knows the proper action to take in the proper context. Still, both intellectual and practical virtues are means to the attainment of happiness, defined as the full realization of the human potential.

Aristotle offers some advice on the details of practical moral judgment. He suggests avoidance of the extremes of excess and defect, defining the virtuous person as one who aims at moderation. Acknowledging that the exact point of moderation varies according to the individual and the context, Aristotle claimed that appropriate action was discerned either through reason or by a right-minded person.

Critics of virtue theory have suggested that universal obligations may be helpful, and even necessary, in certain kinds of human interactions. This is similar to the critique of utilitarianism. Without moral rules, it seems that actions that are obviously wrong can be justified by using virtue theories.

Communitarian Theories

Communitarian ethical theory is most easily understood through a discussion of the philosophy of David Hume, a Scottish historian and philosopher who lived from 1711 to 1776. He asserted that the concept of right and wrong arises from personal sentiment, not from reason. Thus he saw the role of reason in ethics as a tool to discern the consequences of an action already chosen by sentiment. He asserted that the supreme moral good is benevolence, defined as an unselfish regard for the general welfare of society as consistent with individual happiness.

Hume based his moral philosophy on three interconnected propositions. The first is that desire, sentiment, or passion determine one's interests and goals, and reason plays a subordinate role. Hume is famous for his statement, "Reason is the slave of the passion in morals" (Beauchamp 1991, 257), which set him directly against Kant's rationalism. He later explicated this idea further, adding that reason can inform and correct passions but is never the motivating desire to do anything.

Hume's second basic proposition is that people have a natural motivation to act benevolently, but humanity's limited ability to practice benevolence leads society to establish rules of justice. He believed that both natural virtues, inherent in humans, and artificial virtues, those beyond human nature but imposed by the public, are important. Hume agrees with Aristotle that motives alone are virtuous. He sees actions as evidence of motives, but motives as the true objects for appraisal in an ethical theory.

Third, Hume believes that both moral philosophy and social ethics originate in historical traditions and community decisions. This is evidence of Hume's historicism. Moral rules do not arise solely from what an individual feels, but develop within a cultural matrix, where they derive from our common human nature. Hume viewed an ethical system as necessary and convenient for a cooperative community. The goal of morality then becomes the establishment of rules that promote peace and ameliorate conflict.

Later communitarian theory that builds on Hume's original ideas has suggested that morality is needed to address conflicts created by limited resources and sympathies within a community. Both Hume and later communitarians acknowledge that morality is subjective. They accept that what gives rights and obligations their status is their acceptance by the community. This understanding rejects an objective or normative belief in natural law or human rights. Communitarianism is unique in that it allows for moral pluralism along with the acknowledgment that a variety of ethical theories may contain insights, however limited.

Communitarianism is a realistic theory of the origins and purposes of morality, a useful way of understanding development of common morality. One of its important insights is the importance of emotions in morality, but beyond that it does not provide a very useful way to think about real-life ethical problems.

Both the common morality and ethical theories provide standards of conduct or norms, usually phrased as principles or rules. Unless there is a single principle, as in utilitarianism, any application of the theory requires some balancing or relative weighting of these norms. No matter what the preferred moral basis, the principles and rules are usually much too general to apply to real-life situations. Consequently, we need to outline some strategies for using fundamental ideas to resolve practical problems.

Prima Facie and Actual Norms

Moral norms, whether expressed as principles or as rules, are not rigid standards that exclude compromise. Real-life moral problems rarely have clean solutions. Often two rules or principles come into conflict, or the facts of the case make it acceptable to bend or break a rule. For example, truthfulness is an important moral principle, but in many situations telling what is usually called a "white lie" can be justified. The philosopher W. D. Ross has distinguished between *prima facie* and actual obligations (Beauchamp and Childress 2001). A *prima facie* obligation is one that must be fulfilled unless it conflicts with some equal or stronger obligation. A *prima facie* obligation is related to clear cases, those actions that are clearly right or wrong.

When confronted with a situation in which two or more *prima facie* obligations are in conflict, morally responsible people must analyze the competing factors and determine what their *actual* obligation is. An example of a classic moral dilemma that involves competing obligations can be expressed in the question, "Would you lie to save a life?"

The application of rules and principles to real moral problems involves two processes. First, the rule or principle must be specified for the particular situation. Justice, or fairness, is a core moral principle, but in a particular circumstance the principle might be interpreted in several equally valid ways. Suppose a group of people are snowbound and have only a limited supply of food. Does justice require that the food be distributed equally? Should, perhaps, a person who is sick or a small child receive more than a healthy adult? What seem to be strict moral rules have exceptions that need to be specified. Specification, however, cannot solve every problem. Two moral principles or rules might be in direct conflict. Such cases require balancing deciding which principle or rule will be given the greater weight. Morally responsible balancing requires that good reasons be given for the decision, and there are certainly cases in which two people of good character might disagree.

The possibility that morally serious people might come to different solutions to a particular problem does not, however, imply ethical relativism, the idea that moral standards vary from person to person or from culture to culture. Although there are certainly observable differences in patterns of moral judgment between cultures, and even within cultures, this fact does not imply differences in moral standards. In fact, there are good reasons to believe that a widespread agreement on core moral values transcends cultural differences and that essentially all people of good character will agree on the rightness or wrongness of clear cases. The common morality that derives from our human experience does appear to be universal.

That being said, there are communal norms that differ in various ways from common morality. These are the values shared by specific groups including professions. We turn to the subject of professional ethics in the next chapter.

Further Reading

There are a large number of excellent introductory books on moral philosophy. The ones that I have found useful and accessible include Beauchamp (1991), Rachels (1999), and Blackburn (2002).

Chapter 3

Professionalism and Ethics in Chemistry

Common morality and ethical theory are universal. Not only do they provide the standards of conduct that we expect all rational persons to follow, they also provide the basis for professional ethics, the special rules of conduct adhered to by those engaged in pursuits ordinarily called professions, such as law, medicine, engineering, and science. Although common morality and ethical theory are general, professional ethics is specific. Legal ethics applies only to lawyers (and no one else); scientific ethics applies only to scientists. Professional ethics is consistent with common morality, but goes beyond it. Professional ethics governs the interactions among professionals and between professionals and society (Callahan 1988; Bayless 1981). In many cases it requires a higher standard of conduct than is expected of those outside the profession, but the norms of professional ethics must be consistent with common morality. To understand professional ethics, it is necessary to understand the concept of a profession (Davis 1998).

The Concept of a Profession

A profession is more than a group of people engaged in a common occupation for which they are paid. While there are a variety of ways to define a profession, I use a social contract approach, which I have found to be most useful in my thinking about professional ethics. In this view a profession derives from two bargains or contracts: one internal and one external. The internal bargain governs the interactions among members of the profession while the external bargain defines the relationship of the profession to society. Both, however, are based on a moral ideal of service around which the profession is organized (Davis 1987). For lawyers the ideal is justice under law. For physicians the ideal is curing the sick, protecting patients from disease, and easing the pain of the dying. As Michael Davis has argued, these moral ideals go beyond the demands of ordinary morality, the requirements of law, and the pressures of the market. Using a moral ideal as the fundamental basis of the profession comes from the old-fashioned idea of a profession as a calling. If members of a profession share a moral ideal, then the internal code of practice and code of ethics that develop out of that ideal have an authority that supersedes mere social convention or fear of sanctions; they represent the core values of the profession.

For client-oriented service professions, such as law and medicine, the underlying moral ideal is fairly easy to identify. But is there a moral ideal for chemistry or for science in general? Simple as this question seems, it is complicated because science is not a monolith. The disciplinary and professional culture of chemistry is quite different from that of physics or biology. Even within chemistry there are significant differences between the perspectives of organic and physical chemists. While I don't yet have a complete answer, I have identified two important parts of a moral ideal for science: the habit of truth and the ideal of the gift economy.

A Moral Ideal for Science

A powerful statement of at least part of the moral ideal of science can be found in Jacob Bronowski's book, *Science and Human Values* (1956): he calls it the *habit of truth*. Science is the dispassionate search for the understanding of nature, what John Ziman (1978) has called "reliable knowledge." Further, scientific truth is considered to be of intrinsic value, independent of its applicability. Although science does lead to useful products and inventions, such applications are only secondary to the search for what Einstein called "the secrets of the old one" (French 1979, 275). The best scientific research is driven by an insatiable curiosity about the way the world works. And because scientific knowledge is severely constrained by experiment, scientists are bound by what Richard Feynman (1985, 341) called "a principle of scientific thought that corresponds to a kind of utter honesty—a kind of leaning over backwards." Gerald Holton (1994), quoting P. W. Bridgman, called it "doing your damnedest, no holds barred."

The second part of the moral ideal concerns the relationships between scientists: the principle of the gift economy. Because scientific research is so difficult and because science is public knowledge, the scientific community is bound by an ideal of relationship and open communication exemplified by the gift economy (Hyde 1979).

The concept of a gift economy is best introduced by contrasting it with the commodity economy, which governs our day-to-day economic interactions. Transactions in the commodity economy are mutually beneficial, closed interactions: fee for goods, fee for service. We go to the grocery store and buy a quart of milk for the listed price, and both parties are happy. No further relationship (except perhaps that governed by a warranty) between buyer and seller is expected or desired.

On the other hand, the gift economy is characterized by open interactions: people give each other advice; they do favors for each other; they coach and referee children's sports. Gift economies serve to bind people together and create mutual obligation. Commodity economies work under fairly strict rules that define and delimit mutual responsibilities and future obligations between the parties involved. Gift economies aim to initiate and maintain human interactions. One becomes a part of the gift economy by contributing something, by giving a gift. In the gift economy, those who are valued most are those who give the most. In the commodity economy, the most important people are those who have accumulated the most.

Pure science operates largely as a gift economy (Baird 1997; McSherry 2001). Scientists contribute their work and often a great deal of their time without any specific expectation of a financial return. They contribute intellectual and creative gifts to the community in the form of their research results: experimental procedures, data, interpretations, and theories, for example. They contribute their time in presenting the results of their research at other institutions and at professional meetings without compensation, except perhaps for travel expenses and, in the best circumstances, a modest honorarium. Likewise, they referee articles and grant proposals. Most of the essential peer review process in science is part of the gift economy. Some serve as editors of journals and books, again with little if any financial compensation. They receive in return similar gifts from other members of the scientific community, but there is no quid pro quo.

To be a member of the scientific community therefore, one must contribute. The greatest scientists are those who contribute most, particularly in quality of work. For example, Linus Pauling was one of the greatest chemists of the twentieth century because his insights into the nature of chemical bonding, which he freely presented to the world, are used daily by working scientists; his findings redefined chemistry. On the other hand, Thomas Alva Edison developed and adapted scientific discoveries into salable commodities from which he gained profit, but he gave nothing back to the scientific community and, in fact, earned the ire of Henry Rowland who once complained that the "spark of Faraday blazes at every street corner" (Moore 1982). Rowland felt that Michael Faraday, who made the fundamental discoveries, should get the credit, not Edison, who merely developed the commercial product. While Edison contributed much to society through his inventions, he was not really considered a part of the scientific community because he contributed little either in fundamental knowledge or in experimental or theoretical techniques.

While moral ideals may be the basis, professions are complex social organizations that evolve over time. The classic learned professions—law, medicine, and church—have long histories, though both medicine and law began to take their modern form only in the nineteenth century as improved transportation and communication fostered communities of practitioners (Starr 1982). Chemistry has a comparable history (Knight 1992; Brock 1992).

The Development of Professions

Professions develop through a historical process of self-definition. Although chemistry is an ancient science that began with the crafts of metal working, ceramics, dyeing, and tanning, it really began to define itself as an independent science in the eighteenth century, particularly with the work of Lavoisier, and matured to a fully recognized independent field by the middle of the nineteenth century when chemists began to organize scientific societies (Knight and Kragh 1998). The American Chemical Society (ACS) was founded in 1876 by a group of New York chemists who had attended the first American national meeting of chemists in Northumberland, Pennsylvania, held in 1874 to commemorate Joseph Priestley's 1774 isolation and characterization of oxygen (Reese 1976). The ACS was originally founded as a scientific rather than a professional society. Although some thought was also given to student training and improving the public image of chemistry, its purpose was mainly to encourage research by holding scientific meetings and publishing journals. Professionalism became an explicit concern of the ACS in the 1930s largely due to the economic pressures of the Great Depression. Partly because of the large number of chemists employed by private industry, professionalism is a continuing concern of the ACS. In fact, there is internal tension between the interests of industrial and academic scientists.

Similar organizations of chemists were founded in Europe in the nineteenth century. In England, The Chemical Society was founded in London in 1841 to bring together academic, manufacturing, and consulting chemists. The Faraday Society, which was to be a bridge between science and technology, particularly electrochemistry, arose in 1902. In 1857 a group of junior chemists in Paris formed what became by 1859 the Société Chimique. The Deutsche Chemische Gesellshaft was founded on the British model in Berlin in 1867 (Knight and Kragh 1998). By 1900 chemists internationally were well organized as were the other physical sciences. The founding of chemical societies signifies the establishment of a disciplinary and professional identity though the development of an internal code of practice.

Internal Code of Practice

As part of the process of self-definition, members of a profession must agree on an internal code of practice and negotiate the relationship between the profession and society. The internal bargain consists of standards of education and training; a formal or informal certification or licensing procedure; and a code of practice, which usually includes a formal code of ethics. Some professions such as law, medicine, and engineering have well-defined standards of education enforced by accreditation boards. In science, the standards are less formal, although chemistry, at least in the United States, is unique in having developed formal standards for undergraduate education. The American Chemical Society has a Committee on Professional Training, which establishes standards for a professional baccalaureate degree and approves the curricula of individual institutions (Committee on Professional Training 1999).

The standards of training in science have evolved over the years, but currently an earned doctorate from a reputable university is the usual requirement. Scientists without doctorates can, however, be recognized after publication of credible research in refereed journals. For example, Charles Pedersen, an industrial chemist with a master's degree, won the Nobel Prize for his work on crown ethers. Formal certification is not common in the physical sciences, so recognition comes from accomplishments rather than a professional license.

There have been a number of attempts to formulate the internal code of practice of science. Perhaps the most famous is that of Robert K. Merton (1973). Merton identified four principles of scientific practice:

1.   *Universalism*: Truth claims must be evaluated using preestablished impersonal criteria.

2.   *Communism (or "communality")*: Scientific findings must be disclosed publicly. In Ziman's (1968) terminology, science is public knowledge.

3.   *Disinterestedness*: The advancement of science is more important than the personal interests of the individual scientist.

4.   *Organized skepticism*: All scientific truth is provisional and must be judged based only on the evidence at hand. Scientific conclusions are always open to question. [This is similar to Popper's (1965) famous principle of falsifiability.]

Merton's list has been modified and expanded by later writers to include such ideas as objectivity, honesty, tolerance, doubt of certitude, selflessness, individualism, rationality, and emotional neutrality (Barber 1952; Cournand and Meyer 1976; Zuckerman 1977). Although Merton viewed these four principles as merely descriptive of the practice of science, they have ethical implications. It is easy to see their relationship to the two parts of the moral ideal of science: the habit of truth and the ideal of the gift economy. Universalism, disinterestedness, and organized skepticism are integral to the pursuit of reliable scientific truth, and "communism" is an aspect of the gift economy.

Merton's principles are analogues of the broad moral principles, like justice, mentioned in Chapter 2. They are the basis of more specific moral rules that govern the day-to-day practice of science. Some of these more practical rules are

1.   Experimental and theoretical procedures are reported accurately so that independent investigators can replicate the work if they so choose.

2.   The data reported are complete and correct, and the limits of error are also noted. Scientists are not supposed to suppress data that do not agree with their expectations.

3.   The interpretation of the data is done objectively. Prior expectations should not interfere with data analysis, and nonscientific factors, such as politics or the expectations of the funding agency, should not influence the analysis.

4.   Credit is given where credit is due. Scientists are expected to cite previous work where appropriate and to give credit to those who have aided in the research. Conversely, it is assumed that all the authors of a scientific article have contributed to the research.

David B. Resnik has compiled a longer list of the standards of ethical conduct in science in his recent book, *The Ethics of Science: An Introduction* (1999). Resnik's standards include those listed above, although he states them differently, as well as norms regarding the education of future scientists, the freedom to pursue research topics, social responsibility of science, the opportunity to use resources, efficiency of resource use, the treatment of living research subjects, mutual respect, and legality.

In addition to these norms there are more specific research practices that can vary depending on the discipline. Some of the criteria that distinguish "good physics" are different in kind from those that distinguish "good chemistry." Learning the techniques and standards of research in a particular discipline is a major part of the graduate education of a scientist. It is what Thomas Kuhn (1962) has called normal science.

Along with ethical norms, the internal code of practice of any profession or discipline also contains standards of etiquette, though the two often overlap. Rules of etiquette govern behavior but lack moral content. For example, questions of coauthorship involve both ethics and etiquette. While it is ethically important that everyone who contributes to a scientific paper receive credit, the order of the authors is largely a matter of etiquette. In fact, a variety of conventions govern the order of authors. Sometimes the senior (or most important) author is first; sometimes last. As an undergraduate I was told that "gentlemen publish alphabetically." This was before the use of gender-sensitive language was common. I even once read a paper that contained a footnote stating that the order of the authors was determined in a poker game. The scientists who maintain large instrumental facilities, such as neutron sources, are included as coauthors as a matter of etiquette on articles containing data obtained using these facilities, but technicians who work on research projects are often only given credit in the acknowledgments. Learning these conventions is also a part of scientific training.

It is important to recognize that the internal code of science has evolved over time. While the broad principles of the code go back to the early days of the Royal Society of London, specific details and norms of scientific practice have changed significantly since Boyle and Newton. Therefore, recent charges of scientific fraud directed at historical personages should be regarded with some skepticism (Holton 1994).

Both the American Chemical Society and the American Institute of Chemists have developed formal codes of ethics. In addition, the American Chemical Society has adopted a detailed set of "Ethical Guidelines for the Publication of Chemical Research." These codes of ethics are formal statements of the internal and external bargains of the discipline. (See the Appendix.) They also have a more profound significance.

Epistemology and Ethics

On one level the internal code of practice of a science can be regarded merely as a social convention. On a deeper level, however, it has both ethical and epistemological significance. In several articles, John Hardwig (1985, 1991, 1994) has argued that in science (and other fields) epistemology and ethics are intimately intertwined. As a chemist, I claim to know things about chemistry and other sciences which I have not thoroughly studied myself. In some cases, I have neither the background nor the ability to follow the detailed arguments that established the knowledge I claim as my own. As Hardwig points out, I believe many things are true merely because I trust that the scientists who report them did the appropriate experiments or theoretical calculations and interpreted them correctly; I accept their testimony as truth. Therefore, my knowledge depends on the moral character of other scientists. While I think all scientists tacitly understand that our knowledge depends heavily on the integrity of others, this issue is rarely discussed.

Hardwig's analysis (1991) is based on the idea of epistemic dependence. As scientists depend on each other for knowledge, an unequal power relationship develops; one person becomes the "expert," the other a "layperson." Each has ethical responsibilities. Experts must be careful in what they say, and laypersons must be careful in evaluating and using the information they receive. Further, the community of experts has ethical responsibilities to ensure that its members behave responsibly. Hardwig presents a preliminary set of maxims to clarify the ethics of expertise. Many of these maxims are consistent with the internal code of practice of science, while others address the appropriate relationship between science and society. From my perspective, probably the most important maxim for the community of experts is "Take steps to ensure that your members are worthy of the social trust placed in them." This maxim captures the essence of the bond of trust that should exist within the scientific community and between science and society. Steven Shapin has traced the development of the role of trust in science in *The Social History of Truth* (1994).

Beyond the social aspects of trust in science, there is the question of trusting oneself. As chemist and philosopher Michael Polanyi (1964, 70–71) has so nicely explained, the acquisition of knowledge is a skillful act of personal commitment, the art of knowing: "To affirm anything implies, then, to this extent an appraisal of our own art of knowing, and the establishment of truth becomes decisively dependent on a set of personal criteria of our own which cannot be formally defined." Although Polanyi does not explicitly say so, among those personal criteria must be ethical standards. Failure to apply personal ethical standards to the act of knowing results in the kind of self-delusion that led, in part, to spurious claims of N-rays and cold fusion (Gratzer 2000). One of Hardwig's maxims speaks directly to this issue: "Know your own ethical limits." In my view, the ethics of personal knowledge is essential to the moral structure of science.

The internal bargain provides the norms for professionals in their work within the professional community. Yet no professional community exists in isolation, so we must turn to a discussion of the external bargain.

External Bargain: Science and Society

The external bargain addresses the relationship of the profession to society. In general, the profession lays claim to a body of specialized knowledge and skill not easily attainable by most people. In return for a monopoly on the practice of those skills, the profession agrees to use them to serve society and to render professional judgment when asked. For professions such as law, medicine, and engineering, the bargain with society is highly structured; parts are even written into law. For science the agreement is less formal. A brief historical sketch will help to clarify the relationship of science and society.

Perhaps the first agreement between science and government came with Charles II's establishment of the Royal Society of London. The Royal Society was given the right to publish without censorship and pursue the new specialty of natural philosophy. In return, the Royal Society was to avoid the study of politics, morality, and religion. In the words of Robert Hooke the "Business and Design" of the society was "To improve the knowledge of natural things, and all useful Arts, Manufactures, Mechanics, Practices, Engynes, and Inventions by Experiments (not meddling with Divinity, Metaphysics, Moralls, Politicks, Grammar, Rhetoric or Logick)" (Proctor 1991, 33). The gentlemen who founded the Royal Society established the early standards for scientific practice. The central question was, What should be considered scientific truth? Robert Boyle was a central figure in this development (Shapin 1994).

In the early Royal Society an interesting tension between what we would now call pure and applied science can be exemplified by Boyle and Hooke. While Boyle was the paradigm of the Christian gentleman scientist, Hooke, the curator of experiments, was considered the "greatest mechanick this day in the world" (Shapin 1989). Boyle's disinterested gentility contrasted sharply with Hooke's protection of patent rights. Although both men were more complex than this polarized comparison suggests, the division between pure and applied science is a key issue in science policy, and the tension between the scientific ideal of open communication of the gift economy and the personal economic gain of the commodity economy is an important contemporary issue in professional ethics in science (Kovac 2001; Coppola 2001).

Although the practical aspects of science have always been important, the so-called German model of pure research dominated the development of science in the United States (Reuben 1996). Science in the universities and research institutes was pursued for its own sake. Practical applications were certain to follow as the secrets of nature were revealed. American science could point to outstanding examples of the practical utility of pure science, for example, the work of Nobel laureates Irving Langmuir at General Electric on the improvement of incandescent lights and John Bardeen at Bell Laboratories, who was co-inventor of the transistor.

World War II changed the nature of research in America forever. The Manhattan Project and the development of radar showed how science, with generous government support, could make significant accomplishments in a short time. The new bargain between science and society was outlined in two postwar reports: Vannevar Bush's *Science: The Endless Frontier* (1990) and John R. Steelman's *Science and Public Policy* (1947). These reports led to our current system of research funding centered on the National Science Foundation and the National Institutes of Health.

The essence of this bargain can be summarized in a few words: "Government promises to fund the basic science that peer reviewers find most worthy of support, and scientists promise in return that the research will be performed well and honestly and will provide a steady stream of discoveries that can be translated into new products, medicines or weapons" (Guston and Kenniston 1994, 2). The postwar bargain summarized in this quotation, which characterized scientific practice in the United States through most of the last half of the twentieth century, began to unravel in the 1990s, in part because federal funding was unable to keep pace with scientific development but also because of an increasing emphasis on commercializing the products of research (Brown 1992; Guston 1999). To understand these recent developments, we need to look more closely at the traditional distinction between pure and applied research.

Pure and Applied Research

As Donald Stokes has pointed out in his recent book, *Pasteur's Quadrant* (1997), a simple bipolar classification of research does not really describe the practice of science. Instead of placing the various kinds of research along a linear scale ranging between pure and applied, Stokes proposes the quadrant model shown in Figure 1. Along one axis the research is classified in terms of the quest for fundamental understanding. Some research, such as that concerned with the deep meanings of the quantum theory, is entirely focused on fundamental scientific understanding, whereas much of what we usually call applied research merely uses the results of fundamental research for practical purposes. Stokes's great insight was to recognize that there is a second axis: consideration of use. Although fundamental research in cosmology, say, really has no immediate uses, other quite fundamental research is motivated by practical considerations. Stokes uses Pasteur as his paradigm. While Pasteur made many fundamental contributions to the developing science of microbiology, much of his work was motivated by the practical problems of French industry.

Stokes's scheme results in four broad categories of research, three of which he has named after famous people. The pure fundamental research done in Bohr's quadrant is the "science" philosophers of science usually consider; much of the science done in universities falls into this category. Edison's quadrant, the realm of applied research, is also familiar as the research done in industrial laboratories applying fundamental principles to the development of useful products and processes. This is the chemistry of dyes and personal care products, the legacy of the craft tradition. The interesting quadrant is Pasteur's: use-inspired basic research. Although Stokes uses Pasteur to illustrate, much of chemical research lies squarely in this quadrant.

An excellent example is the work of Wallace Hume Carothers on condensation polymers at DuPont. Although Carothers was looking for a way to demonstrate the existence of macromolecules using well-known organic reactions, a controversial hypothesis at the time, his work was inspired

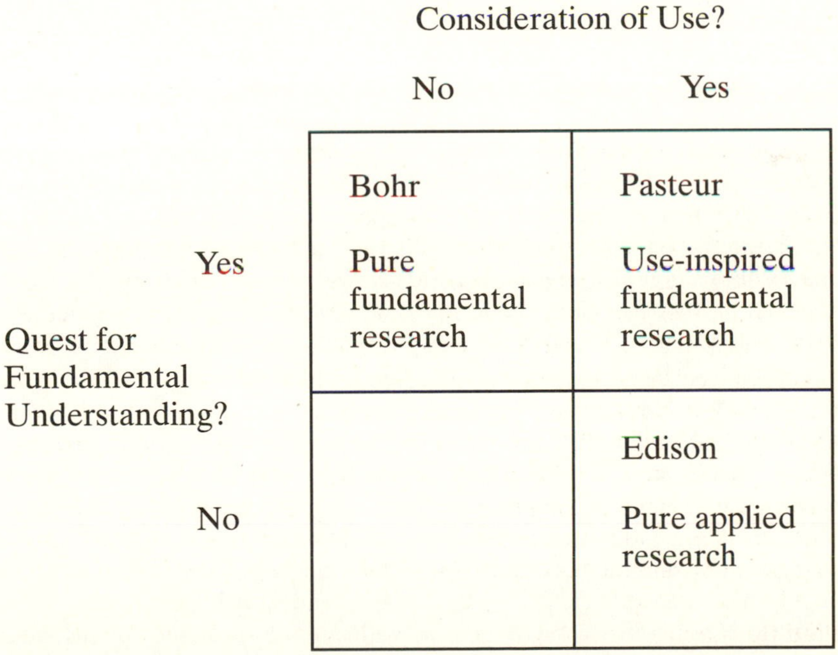

Figure 1   Quadrant model of research. *Source*: Derived from Stokes 1997.

by the prospect of developing commercial products. In the process, he invented nylon, the first synthetic fiber (Hermes 1996; Hounshell and Smith 1988; McGrayne 2002).

One of the distinctive features of chemistry as a science is its emphasis on synthesis, creating new substances (Rosenfeld and Bhushan 2000). Synthetic chemists are not only inspired by the challenges of making new molecules and developing new reactions but often also by the possibility that the new molecule will be useful, perhaps as a pharmaceutical or a material. Synthetic chemistry can fit into three quadrants. Some synthetic projects are undertaken just to make theoretically or aesthetically interesting molecules, such as the platonic solid analogues like cubane. Much synthesis, particularly that done in industry, is purely applied research; known reactions are applied to the development of practical products. However, a significant fraction of synthetic chemistry resides in Pasteur's quadrant—and not only because of the pressures of funding. The twin challenges of fundamental research and creating a useful product can be intellectually intoxicating.

As Stokes and others have pointed out, research, even in universities, increasingly operates in Pasteur's quadrant (Guston 1999; Davis 1999). The postwar implied contract between science and society is giving way to a new research policy that emphasizes productivity and technology transfer. In addition, the Bayh-Dole legislation that allows universities to patent the results of federally sponsored research has led to an explosion of entrepreneurial activity by university scientists, particularly in the biological sciences, but also in chemistry (Coppola 2001). Faculty are being encouraged both to patent their discoveries and to start their own companies to commercialize the products or to license them to existing companies. In addition, industry is coming to universities for basic research, and faculty and administrators are delighted to accept the contracts to replace dwindling federal support. When the scientific results become commodities, something is gained by society. But science suffers an important loss because of the possible tensions with the moral ideal of the gift economy (Kovac 2001).

Trust in Science

Both the internal and external bargains, like all human bargains, depend crucially on trust. As noted earlier, members of the profession are epistemically dependent on each other so they must trust each other to follow the professional code. Because society depends on scientists as experts in important personal and public decisions, it must trust that scientists are performing their work with integrity, particularly as it affects public health and safety. Arnold S. Relman, editor of the *New England Journal of Medicine*, states the importance of trust in science particularly well: "It seems paradoxical that scientific research, in many ways the most questioning and skeptical of human activities, should be dependent on personal trust. But the fact is that without trust the research enterprise could not function" (quoted in Djerassi 1991).

Trust among scientists and between science and society has eroded over the past ten years. Trust between science and society is eroded more these days by science's changing its position on issues that the public cares about, particularly in the biomedical sciences. Debates among scientists about the efficacy of drug therapies, product recalls, and the like because of "inadequate science" in the testing are much in the news. Reports of real or alleged scientific misconduct appear regularly in the popular and scientific press (Broad and Wade 1982; LaFollette 1992; Bell 1992; Zurer 1987). While there is controversy over how widespread ethical problems in science are (Swazey et al. 1993; Goodstein 1991), concern in the scientific community is significant. In 1992 the National Academy of Sciences issued a comprehensive report entitled *Responsible Science:*

Ensuring the Integrity of the Research Process (Panel on Scientific Responsibility and the Conduct of Research 1993). This report contains a number of recommendations to ensure integrity in the research process. For our purposes here the most important is recommendation 2: "Scientists and research institutions should integrate into their curricula educational programs that foster faculty and student awareness of concerns related to the integrity of the research process" (Panel on Scientific Responsibility and the Conduct of Research 1992, 13). In other words, we need to teach scientific ethics.

Teaching and Learning Scientific Ethics

Traditionally, scientific ethics has been taught informally by example or in the context of actual issues that arise in the conduct of research. As Richard Feynman (1985, 341) noted in his Caltech commencement address, "That is the idea we all hope you have learned in studying science in school—we never explicitly say what this *is*, but just hope you catch on by all the examples of scientific investigation. It is interesting, therefore, to bring it out now and speak of it explicitly." In the smaller, slower-paced scientific enterprise of earlier times, this informal method worked well. Contemporary science is bigger, faster-paced, and filled with more complicated issues. Economic pressures are enormous. Research groups are larger, so faculty advisors and students have less personal contact. Contemporary science is expensive, so the grant-writing process is hugely important and time-consuming. In addition, opportunities and pressures for commercializing research are increasing, particularly in the emerging biotechnology industry. Because the scientific workforce is much more diverse, it is not safe to assume that all scientists share a common value base. Media and government scrutiny of science is intense. All these factors call for a systematic approach to the teaching and learning of scientific ethics (Davis 1993; Committee on Assessing Integrity in Research Environments 2002; Coppola and Smith 1966; Coppola 2000).

When I discuss the teaching of scientific ethics with my colleagues and with the public, I hear two major objections: (1) professional ethics is best taught in the research group and (2) we can't teach ethics; either people are moral or they are not. To the first objection I would respond that the research group may be the best place to teach scientific ethics, but that one cannot guarantee either that it will be taught there or that it will be taught well. I was lucky to have research advisors and mentors who had high standards of professional conduct and provided me with excellent role models. Not everyone is so fortunate. Such an important aspect of professional education cannot be left to chance. The second objection reveals a confusion between ordinary morality and professional ethics. I would agree that instruction in professional ethics cannot transform a fundamentally immoral person. On the other hand, even if students come to college as fairly sophisticated moral decision makers in their day-to-day lives, they probably are not skilled at making decisions regarding professional ethics. The two are different.

Science is filled with ethical decisions. Many decisions that on the surface seem purely technical also involve professional ethics. Some examples include

1. *Discarding a data point*: Data points are discarded in many experimental investigations for legitimate reasons, such as contamination of the sample, improper functioning of an instrument, or procedural errors. An important component of professional judgment in science is the ability to know when an experiment has worked properly. At the frontiers of research the signal may not be easily distinguishable from the noise (Holton 1978). On the other hand, there are a number of historical examples of scientists whose expectations of the outcome of

an investigation caused them to misinterpret data, either discarding relevant data or retaining incorrect measurements (Franks 1981; Close 1991; Nye 1980; Gratzer 2000).

2. *Writing a scientific article or report*: The scientific article is not a dispassionate recording of the details of an investigation; it is an argument (Medawar 1964; Hoffmann 1988; Locke 1992; Gross 1996). An author is faced with many decisions. Who should be a coauthor? Which results should be included? How should they be presented? How should the inevitable "loose ends" be treated? How should potential objections be met? What prior work should be cited? These are all questions in professional ethics.

3. *Laboratory practices and safety*: Experimental science is full of danger. A working scientist has to make decisions concerning the potential hazards of a particular experiment and adopt reasonable safety precautions. Are the potential health and safety risks of a laboratory procedure acceptable? These decisions have an ethical component.

I have only given three examples; there are many others. Questions of professional ethics arise quite naturally both in science courses and in the practice of research. Often, these are treated only as technical issues; the ethical component is ignored. It is important for chemists (and all scientists) to broaden their perspectives and explicitly recognize that most decisions in science involve both technical and ethical issues.

Ethical decision making involves at least four components (Davis 1995; Kovac and Coppola 2000):

1. The ability to identify and articulate the moral dimensions of situations and events including the recognition that many decisions involve both a technical and a moral component.

2. Understanding the relevant standards and ideals that govern moral decisions in professional situations.

3. Awareness of the moral complexity of real-world situations. As has been nicely stated by Caroline Whitbeck (1996), many practical ethical problems call for coping rather than solving. Problems in practical ethics rarely have clean solutions; usually, some moral principle or rule is compromised. Learning to see priorities for what they are, to balance them, and to design solutions to complex moral problems should be a major goal of college education.

The first three are passive skills that are incomplete without a fourth, active, characteristic.

4. Moral courage—the willingness to make difficult decisions, act on them, and publicly state those decisions and the reasons for them.

While discussions of principles and rules provide a context, professional ethics is best discussed in the context of real-life ethical problems where the broad principles and rules must be specified and balanced to find a practical solution. Learning ethics is a lot like learning science. Theoretical knowledge only becomes real when it is applied to specific problems. Most chemical concepts are empirical generalizations. Abstract ideas, such as acids and bases, become meaningful examples as their properties and reactions are studied in the laboratory. The remainder of this book is devoted to practical reasoning about ethical problems or cases.

Chapter 4

Ethical Problem Solving

An ethical problem is not like a mathematics problem or most science problems that have unique solutions that are either right or wrong. Instead, ethics problems are more like design problems for which several acceptable solutions can be found. Design problems are problems of making or repairing things or processes that satisfy human desires or needs (Whitbeck 1996). The most familiar example in chemistry is design of a synthesis, an example of process design. There is usually more than one way to make a particular molecule. Deciding on which method is "best" involves a large number of considerations including cost of materials, yield, quantity and purity of product, safety, purification methods, and reaction conditions, among others. Two different chemists might choose two different routes based on individual considerations. For example, while one route might provide a higher yield but require an expensive piece of equipment, the second route has a lower yield but can be done less expensively. The chemist who already owns the specialized equipment will probably choose the first alternative, but a colleague whose research budget is limited might accept the lower yield to save money.

In a second kind of synthesis design problem the end use is known, but several molecules or materials might actually accomplish this goal. Drug design is a good example. A chemist might take on (or be assigned) the task of developing a compound that controls blood pressure by blocking an enzyme that constricts blood vessels. A number of compounds might work, and the "best" solution to the problem will depend on factors such as ease of synthesis and purification, cost, medical side effects, and safety and environmental considerations involved in the manufacture of the drug.

In general, the design's success depends on whether it achieves the desired end within the imposed criteria and constraints. There is a close analogy between design problems and real-life ethical problems. In an ethical problem a chemist or chemistry student must devise possible courses of action, evaluate them, and then decide what to do. As in a design problem, ethical decisions are often made with uncertain information under a time constraint. The ethical problems that we face are practical problems; an acceptable solution must be found because, in most cases, doing nothing has both moral and practical consequences.

Suppose I am a graduate student and my research advisor instructs me to discard a data point that I think is both valid and relevant to the investigation before analyzing the results. I know that reporting all the data is one of the core principles of science, so this action is a violation of professional ethics. While the ethical principle is clear, the practical question is what I can and should do in the circumstance. I have been given an order and I need to respond. Should I obey the order, ignore it, refuse it, report it to someone? To whom? Another member of the faculty? The department chair? The dean? The research office? Should I do something else? Is there somewhere I can go for advice? What are the consequences of these different actions? Where can I go to find out? There are myriad questions to answer, and they probably must be answered quickly because my advisor will soon be asking for the results of my data analysis.

This simple example shows that ethical problem solving, or ethical decision making if you prefer, is a complex process. While there is no simple algorithm for ethical problem solving, a systematic four-step approach can be outlined. Briefly, the four steps are:

1. *Define the problem*: Collect as much factual information as possible and identify the ethical issues involved.
2. *Collect data*: Determine the alternatives, the parties involved, and the relations among the parties.
3. *Analyze the data*: Assess how each of the alternatives affects each of the parties involved.
4. *Make judgments*: Use moral reasoning to determine which of the alternatives are acceptable, which are unacceptable, and, finally, which is the best course of action.

The four steps can and should be used iteratively, expanding the aspects of the problem, increasing the number of parties, and expanding the number of potential alternatives to examine. Eventually, the analysis should stabilize; all the possibilities will be generated. We now examine each of the steps in more detail.

Definition

A clear statement of the problem is always the best place to start. It is essential to understand the factual situation in as much detail as possible. Answering the five basic journalistic questions—who, what, when, where, and why—is an excellent way to clarify the facts. Second, the important ethical issues must be identified. Which norms of professional ethics might have been compromised? Have general moral principles been violated? Are the rights of individuals being respected? Are there broader implications for society? Asking questions like these will help identify the ethical issues.

Data Collection

Once the situation is clarified, the process of devising alternatives can begin. At this point an ethical problem most resembles a design problem. Brainstorming is a useful technique to generate a large number of possibilities. In this creative process the problem solver does not evaluate the different possibilities but just collects them. Each alternative will involve a number of parties, or stakeholders, that will be affected in some way by the course of action being considered: people, organizations, the scientific community, society at large, even the scientific process itself.

It is also important to identify the relationships among the various parties. In the example discussed above, the graduate student and the research advisor have an important relationship that must be considered. To use a scientific analogy, the couplings between the parties must be identified and their relative strengths evaluated. Some important factors to consider in understanding the relationships include power, obligations, and rights. For example, while research advisors have unequal power relationships with students, they also have certain obligations to their students that are part of the professional code of ethics of chemists.

Data Analysis

To analyze the data, the scientists assess how each of the alternatives affects each of the parties. They determine whether the alternative improves a party's lot or makes it worse. Does the proposed course of action strengthen or weaken the various obligations of the parties? Are fundamental rights being violated?

Judgments

In the final step, each alternative must be evaluated using moral arguments based on the various approaches outlined in Chapter 2. A deontological analysis focuses on the rightness and wrongness of the actions. Principles and rules are invoked. In the case in which the graduate student must decide how to respond to the research advisor's order to suppress a data point, suppressing the data point violates the principle of proper reporting of data. Of course, the principles and rules need to be specified for the particular circumstance and balanced if two or more come into conflict.

In contrast, a consequentialist, or utilitarian, analysis looks at how each party is affected. Does the proposed alternative maximize good or, at worst, minimize evil? In the example, one could argue that suppressing the data point will result in a stronger-looking article which will enhance the reputation of both the advisor and the student, which in turn might result in increased grant funding and other benefits. On the other hand, the scientific community is diminished by having tarnished results published in the open literature, and the public is ill-served by scientific misconduct.

Virtue ethics focuses on character issues. Do the actions of the individuals reflect positive ethical character traits? In our example, a scientist with integrity is not supposed to "cook" data, nor should a research advisor encourage a student to engage in improper behavior. A scientist is supposed to practice the "habit of truth." Finally, a social contract approach is sensitive to the history and values of both the professional and broader community. An evaluation of the situation in the example from this perspective focuses on the importance of trust in maintaining the internal bargain of the scientific community. Suppressing the data point violates an essential principle that is universally accepted by the scientific community. If working scientists cannot trust the results published in the open literature, research cannot proceed. Further, the unwritten contract between science and society has been betrayed. Society relies on science for reliable answers to technical questions.

Approaching the analysis of a problem from multiple perspectives is quite common in chemistry. For example, in understanding chemical bonding, chemists will use experimental data, the simple Lewis model, VSEPR theory, valence bond theory and hybrid orbitals, simple molecular orbital theory, and detailed electronic structure calculations. Each approaches the problem from a slightly different angle. Similarly, each of the ethical theories gives a different view of a moral problem. The deontological approach is top down or deductive, arguing from principles to applications. Consequentialism is more of a bottom up approach. Virtue ethics starts with the agents, and social contract approaches focus on the community. Each is valuable, but incomplete.

With several different methods of analysis, it can be difficult to arrive at a final decision, but the analogy between design problems and ethical problems is instructive. In the best cases, the methods will agree, and one alternative will emerge as the best. In others, the agent must try to sort out which alternative or alternatives seem to be the most viable. Another possibility is that the analysis might point to a new alternative that combines the best features of several good possibilities.

In a design problem there may not be a uniquely correct solution, but there are sometimes clearly unacceptable possibilities. There are wrong answers even if there is no clear right answer. This is also true of ethical problems. Some alternatives are clearly wrong. Either they violate a deeply held fundamental principle, or they lead to poor consequences, or both. One of the objectives of the analysis in step four is to identify the obviously immoral courses of action and reject them.

In interesting and substantive design problems there are usually several attractive solutions that have different kinds of advantages and disadvantages. For example, two materials might work for a particular application. One might be cheaper but have a shorter life span; the other, more expensive but longer lasting. Which is better? The same dilemma arises in ethics. A consequentialist analysis might reveal that two or more alternatives have a similar overall balance of good over evil but with the benefits differently distributed among the stakeholders. The distributions might also be such that even a principle of justice cannot give an unambiguous answer. A deontological analysis may find that it is impossible to find a course of action that does not bend or break a moral principle or rule. There may not be a single right thing to do.

In this context, it is important to note that difficult ethical problems, like difficult design problems, often call for coping rather than solving. Every imaginable solution may lead to a bad consequence or to the violation of some principle or rule. Sometimes the best alternative is not perfect.

Another important technique in ethical problem solving is the use of analogy. Chemists often use analogical reasoning. Compound A is an alcohol, so its reactions should be similar to those of compound B, which is also an alcohol. The periodic table is another powerful analogical reasoning tool. In considering ethical problems it is useful to think about analogous situations. Our simple example of data suppression can serve as a paradigm for other cases in which a temptation to discard a point arises. Case-based reasoning of this kind has a long history in ethics and is quite powerful (Jonsen and Toulmin 1988). Sometimes, in difficult cases, people will agree on the correct course of action even when they disagree about the ethical principles that underlie the judgment.

The methodology developed in this chapter is meant to provide a flexible approach to solving ethical problems. The important thing is not to follow the steps blindly but to come to a morally defensible solution. Common sense and conscience are good guides to conduct, but ultimately it is important to be able to provide good reasons for the chosen alternative.

An opportunity to practice this four-step problem-solving method follows in a set of hypothetical cases presenting real-life ethical problems. These cases represent the kinds of ethical problems that a chemistry student or practicing chemist is likely to face and illustrate the difficulties that can arise in devising a morally acceptable course of action in a complex situation. Each case, or set of related cases, is followed by a commentary. Some of the commentaries are quite detailed and provide careful analyses of the problems. Others are briefer and merely outline the important issues. Just as in the study of chemistry where solving the end-of-chapter problems solidifies understanding, a thoughtful consideration of the cases will bring the abstract principles to life.

Further Reading

Some useful references on ethical problem solving include Brebeau et al. (1995); Harris, Pritchard, and Rabins (1995); Johnson and Nissenbaum (1995); and Jonsen and Toulmin (1988).

Chapter 5

Cases and Commentaries

Just as in chemistry, the best way to learn ethical problem solving is to confront context-rich, real-life problems (Jonsen and Toulmin 1988; Davis 1999, chap. 8). The broad variety of ethical problems or cases presented here are hypothetical situations, but they represent the kinds of problems working chemists and students face. Cases raising similar ethical questions are grouped together. To reach a diverse audience, I have often written several variations of the same situation. For example, a question might be posed from the perspective of the graduate student in one version and from the perspective of the research director in another. For important issues, I have provided cases that are accessible to undergraduates who have very little research experience, usually in the context of laboratory courses. For advanced undergraduates, some cases involve undergraduate research projects. Most of the cases involve situations encountered in graduate research in universities, but some also concern industrial chemistry. Finally, a few cases present ethical problems that arise in cooperative learning, a pedagogical technique that is becoming increasingly important in undergraduate education.

Each case, or related set of cases, is followed by a commentary, which outlines the important issues and discusses possible solutions. Some of the commentaries are quite extensive and actually present and defend my preferred course of action; others are brief and merely raise questions that should be considered in designing a solution. The commentaries model the ethical problem-solving method presented in Chapter 4. As I have emphasized repeatedly, most ethical problems do not have clean solutions. Although some courses of action are clearly wrong, there may be several morally acceptable and defensible ways to proceed. Consequently, readers might disagree with my proposed solutions for good reasons. For example, if I have used a consequentialist approach, my assessment of the relative positive and negative weights of the consequences might be challenged, or I simply might have forgotten to consider some factor. Where I have made a definite recommendation, I have given the reasons for my choice and contrasted it with other alternatives.

As indicated in the introduction, the best way to consider cases is in a group so that the different perspectives of the participants can lead to a spirited discussion. The cases can also be used as prompts for reflective essays. I hope that the cases and commentaries will stimulate even the solitary reader to think more deeply about ethical questions in science.

Further Reading

Additional sources for cases include Bebeau et al. (1995): Committee on Science, Engineering and Public Policy (1995); Frase, Barden, and Kovac (1997); Harris, Pritchard, and Rabins (1996); Macrina (1995); and Penslar (1995). Christiensen et al. (1991) provide excellent advice on leading discussions.

CASE: RESEARCH PROPOSAL DEADLINE

Your grant renewal proposal to National Institutes of Health (NIH) is due in a few weeks. The primary basis for the renewal is a series of experiments being conducted by one of your best graduate students. Unfortunately, the results have not yet shown a statistically significant effect of a certain class of compounds on the rates of growth of tumors in mice. You strongly suspect that such an effect will eventually be proven, but there is not enough time to finish another set of experiments before the proposal deadline. If, however, you omit a few extreme data points, the results will appear to be statistically significant, thereby virtually ensuring that your proposal will be funded and your research continued for three more years. If you include all the data points in the renewal proposal, it will be weaker, and the probability of funding will be decreased. What should you do? Consider the following questions:

1. What are the ethical issues raised in this case?

2. Who will be affected by your decision? List all persons or groups who will be affected. In what ways?

3. What moral principles should you use to decide on a proper course of action? If you use several, which one is most important?

4. Is one course of action best? Are several courses of action acceptable? What are they? And how do you decide which is best?

Commentary: Research Proposal Deadline

The major issue here is the ethics of "cooking" or "trimming" the data. There is pressure on you and the student to produce results for the research proposal. Your student's and your careers depend on continued funding. The temptation is to make things come out "correctly" by making a small change in the data. To some, it will seem analogous to a white lie. After all, if the proposal is funded, you will be able to do the additional experiments to decide whether the outlying data points are really significant. Because you really believe that the work is potentially significant, all you are doing is making the best case possible.

This kind of situation comes up regularly in science. The working scientist is often called on to make a judgment as to whether to keep or discard a measurement. There are many legitimate reasons to throw out data points: the instrument was not working properly, the calibration was not done correctly, the solution must have been prepared incorrectly, there is an impurity in the system, and so on. Also, there are statistical criteria, such as the Q test, which can be used to discard outlying data points properly. (The Q test is discussed in most basic analytical chemistry texts.) When is discarding data points good scientific judgment, and when is it wishful thinking? When does wishful thinking become misconduct? These are important questions to consider.

Although it might seem that omitting a few outlying points to make the proposal stronger is a minor problem, this course of action has some serious negative consequences. Because federal research funding is limited, this proposal might receive support that could have gone to a more promising project that was presented honestly. Since scientific dialogue is based on the assumption that everyone is honest, every instance of dishonesty erodes that fragile bond of trust that keeps the scientific community together.

CASE: YIELDS (1)

You, as a professor, and your graduate student have just finished the synthesis and characterization of a new compound and are working on a communication to a major journal reporting the work. While you have made and isolated the compound, you have not yet optimized the synthetic steps, so the final yield is only 10%. From past experience, you know that you probably will be able to improve the yield to at least 50% by refining the procedure. Therefore, when writing the communication, you report the projected yield of 50% rather than the actual figure. After reading the manuscript your student points out that you have reported the yield incorrectly. What is your response?

CASE: YIELDS (2)

You have just finished the synthesis and characterization of a new compound and are working with your research advisor on a communication reporting the work to a major journal. While you have made and isolated the compound, you have not yet optimized the synthetic steps, so the final yield is only 10%. You are convinced that you will eventually be able to improve the yield to at least 50% by refining the procedure. Therefore, when writing the first draft of the communication, you report the projected yield of 50% rather than the actual figure. After reviewing the draft, your advisor asks you whether the yield you reported was what you actually obtained. What is your response?

CASE: YIELDS (3)

You have just completed the synthesis and characterization of a new compound and are writing an article for publication. The compound is quite interesting and you have developed a clever synthetic route, so you expect this paper to attract a lot of attention. Using your "best" synthetic methodology, you have obtained yields of 40%, 45%, 50%, and 70% on four separate preparations. When you write the experimental section, you report the yield as 70%. Is this report legitimate?

Commentary: Yields

These three cases raise the issue of reporting data that are "better" than those actually obtained. This can be done by "cooking" (manipulation or smoothing), "trimming" (throwing out data at the extremes), or as in the first variation, reporting data that you expect to get once you refine the experimental procedure. Most scientists have probably had the experience of reading an article and remarking, "They couldn't have obtained data that good."

In the case of a synthetic yield, the question of what to report is quite interesting. Should one report the ideal or best yield, the one that can be obtained if everything goes right, or a typical yield or a range of possible yields? With a measurement, say the mass of a substance, we have a good understanding of the random and systematic errors that occur and well-established statistical methods for quantifying them. With a synthesis, there are many sources of error, some of which are hard to quantify. For example, is the product you lost on the sides of the flask a random error? The general ethical principle is the requirement not to mislead, but there are several legitimate ways to specify this norm.

In a sense, a scientific article is an argument and the author is an advocate, a person who tries to put forward the best possible case and to refute all possible objections. Deficiencies will be minimized. On the other hand, the author is also expected to report truthfully the results actually obtained. But a scientific article never reports the actual sequence of events. Rarely does the author mention all the failures, the often multiple attempts to get the experiments to work. Nor does the author report the intellectual groping that usually characterizes the process of understanding the data. Instead, we are presented with a neat logical package, a reconstruction of the actual discovery process. In that context it does not seem unreasonable to report the "best" data rather than the "actual" data. In the situation described in this case, an experienced researcher is convinced based on past history that the yield can and will be improved. Why not report the higher figure? By the time anyone reads the article it will be true.

How honest should you be in a scientific article? That is the essential question. Since much is left out by convention, you are never completely honest. As an advocate you often slant the argument, minimizing doubt, rationalizing small inconsistencies. Much of this is normal science. The question we try to raise in this case is, When have you gone too far?

Two interesting analyses of the nature of a scientific article are found in "Under the Surface of the Chemical Article" by R. Hoffmann (1988) and "Is the Scientific Paper Fraudulent?" by P. B. Medawar (1964). A more extensive discussion of scientific writing can be found in David Locke, *Science in Writing* (1992). For a view of scientific writing from a humanist perspective, see Alan G. Gross, *The Rhetoric of Science* (1996).

CASE: NEARING THE LIMIT

The Saf-Test Corporation is an independent environmental and emissions testing facility that gathers samples from companies and government facilities around the nation and tests them for various chemical and biological agents. Ben works in the water-testing unit, specializing in trace chemicals in lake water. He has done tests on a particular lake near Sinistex Chem for several years, and the concentration of a particular chemical has slowly risen. The acceptable level of this chemical is 5.0 ppm, and the concentration is getting close to this limit. Once this level is reached, the EPA must be contacted to investigate the site.

This month's set of samples has arrived, and the tests indicate problems. The three tests came back as 4.3, 4.2, and 5.1 ppm. Often, the results are similarly scattered, but never before have they been so close to the maximum safe level. When Ben notified his superior of the results, he was told that because the average of the three was below the maximum safe level, nothing should be done. Worried by the one value that was higher than the limit, Ben asked if he could test more samples to make sure that the concentration was safe, but his manager refused to authorize the work, saying that it would cost too much in time and money. What should Ben do?

Commentary: Nearing the Limit

This case contrasts the perspective of the scientist with the perspective of the manager. Scientists practice the "habit of truth." A scientist must consider all the data, rejecting a data point only for good scientific reasons, not because it is inconvenient. A manager wants to keep things running smoothly. For the manager, the outlier that perhaps signals a problem is a danger sign, something to be avoided. These differences in perspective can cause the two to look at the same set of data quite differently.

The manager is able to dismiss the single value that is higher than the limit by pointing to the two values that are less than the limit and the fact that the average is still acceptable. For the scientist, it is important to determine whether the highest value is accurate or whether it is due to some sort of error. The way to check this is to do more tests and make sure the average is still less than the allowed limit.

Ben is also concerned about public safety. The concentration is dangerously close to the maximum. If it really is above the limit, the local population might be subject to a health hazard. At the minimum, the site will be investigated and Sinistex may be required to alter its manufacturing process or to install expensive waste-cleaning equipment. Sinistex management will not be pleased if the reported number is high.

As a professional, Ben's responsibility is to make sure that his analyses are accurate. If that means performing more tests, then he should do them. As an employee, however, he may not have that freedom. The question for Ben is what he judges to be his professional responsibility. If he decides that the manager is not being responsible, then he needs to decide if he should challenge the manager's decision. This is a dangerous step; it could lead to some sort of reprisal. At the extreme, Ben could lose his job.

The tension between the professional (in this case, an engineer) and a manager is discussed by Michael Davis (1998) in an analysis of the *Challenger* disaster (chap. 4).

CASE: DATA POINTS (1)

You are engaged in the study of the substituent effects on the rates of reaction of a particular class of compounds. Based on a variety of considerations you expect that the rate constant will vary linearly with the increasing polarity of the substituent. The graduate student doing the experiments comes in one day with a graph of rate constant versus polarity with the results of ten different systems plotted. Eight of the ten fall nicely on a straight line, but two points are well above the line. You are convinced that the two "deviant" points are in error. What should you do?

1. Do you tell the student to repeat the two "deviant" measurements?
2. Do you tell the student to repeat all the experiments to make sure the data are correct?
3. Do you publish the data omitting the two points that do not fall on a straight line?

CASE: DATA POINTS (2)

You are engaged in the study of the substituent effects on the rates of reaction of a particular class of compounds. Based on a variety of considerations you expect that the rate constant will vary linearly with the increasing polarity of the substituent. The graduate student doing the experiments comes in one day with a graph of rate constant versus polarity with the results of eight different systems plotted. You praise the student's work and ask him to begin writing a rough draft of an article reporting these experiments. He then admits that he actually studied ten systems but left two points off the graph because the data points fell significantly above the straight line. What should you do?

1. Do you tell the student to repeat the two "deviant" measurements?
2. Do you tell the student to repeat all the experiments to make sure the data are correct?
3. Do you publish the data omitting the two points that do not fall on a straight line?

CASE: DATA POINTS (3)

You are a student engaged in the study of the solvent effects on the rates of reaction of a particular class of compounds. Based on several considerations your research advisor expects that the rate constant will vary linearly with the increasing polarity of the solvent. She is quite excited about this project and has been pushing you for results. You have measured the reaction rate in ten solvents, and eight of the ten points fall nicely on a straight line when plotted against a common measure of solvent polarity. The other two, however, are well above the line. You are writing a progress report on this project for your research adviser. Do you include the results of all ten experiments? If so, do you suggest that the two deviant points should be remeasured or perhaps that all ten points be redone? Or do you just tell her about the eight that match her expectations, judging that the other two can be justifiably ignored?

Commentary: Data Points

The preceding three similar cases are concerned with data that do not conform to your expectations. These situations occur frequently in science. Judgments are always being made as to whether a particular measurement should be retained or thrown out. The variety of good reasons for rejecting a data point include instrument malfunction, presence of impurities, and poor technique. Much good science is done right at the edge of the detection limit, so errors are common. The ethical question is whether the data point is being rejected because of good scientific judgment or because of wishful thinking.

The three variations of the case present three alternatives, each of which can be justified. Alternative 2 is the safest: going back and remeasuring everything. But this option might be wasteful, both in time and material. In some circumstances it might be essentially impossible. For example, the measurements might have been performed with an instrument to which you do not have ready access, a synchotron or a neutron beam. Deadlines might be pressing, making this option unattractive.

Alternative 1 can also be justified, depending on your relative confidence in the various measurements. It might be that the two "deviant" points were the hardest systems to work with. They might have been done on a day when the instrument was "acting up." Under these circumstances remeasuring these two points alone can be justified.

Alternative 3 can also be justified under circumstances similar to those used to justify the second. The scientific reasons to doubt the validity of those two measurements may be very good, and it may be impossible or overly expensive to repeat them prior to publication. It might than be perfectly legitimate to throw them out.

After you have chosen one of the three alternatives, what is your responsibility to the scientific community as you write your paper? Must you reveal that you had to remeasure two points or that you threw them out? What should you disclose and what can you ethically leave out?

A useful and interesting reference is G. Holton, "Subelectrons, Presuppositions, and the Millikan-Ehrenhaft Dispute" (1978).

CASE: EXPECTATIONS (1)

As a second-year graduate student in a major chemistry department you are conducting a series of experiments designed to verify your research director's "pet" hypothesis. He expects that the rate constant of a reaction will depend linearly on a particular property of a substituent group. You have carefully studied a large number of reactions, and the data look random to you. But each time that you show him the graph he asserts that he sees a straight line emerging and sends you back into the laboratory to obtain more points. This goes on for months, and it becomes clear to you that the expected correlation just doesn't exist. What do you do? In deciding, consider the following questions:

1.   What are the possible courses of action for the student in this situation? What ethical questions are raised by each?

2.   Who will be affected by each of the possible courses of action?

3.   What moral principles can be used to decide which action is the best?

4.   When you have decided which course of action is the best from an ethical point of view, are there practical considerations that might make this strategy difficult to implement?

5.   To whom might you turn for advice on what to do in this case?

Commentary: Expectations (1)

The student here is faced with the question of whether she should trust her data even though they do not conform to her advisor's hypothesis. There will be a serious temptation to reexamine the data and perhaps throw out the points that do not fit the expected correlation. One can usually find some reason to eliminate a data point, particularly if it does not conform to a theoretical expectation. Another alternative is to "cook" the data so that they conform to the hypothesis. One can remeasure data points and "discover" that they were too high or too low or can find some error in the procedure that justifies discarding the measurement. Seemingly legitimate reasons for adjusting the points can usually be found.

  The alternative is for the student somehow to confront the advisor with the failure of his hypothesis. This can be very difficult. If the student is confident that the data are correct, this is the best alternative, but in some instances it may be difficult to find a good way to approach the advisor. Some advisors are strong-willed and powerful people who easily intimidate students. If the advisor really believes in the hypothesis, he may send the student back to the lab to repeat the experiments, refusing to believe that the data are correct. An inexperienced or meek student might return to the lab determined to find the results her advisor expects, perhaps cooking the data to get the right answer.

  This case is based on an incident described in "Misconduct in Research: It May Be More Widespread Than Chemists Like to Think" by Pamela S. Zurer (1987). In the actual incident, the student showed the advisor a plot produced using random numbers. When confronted with this random data that also fit his hypothesis the advisor admitted to himself and the students that his interpretations had been wishful thinking.

CASE: EXPECTATIONS (2)

One day a senior colleague, Dr. Mavis, comes to you to report an exciting discovery. He tells you that he has been working on a project off and on for the past ten years and has finally discovered a new kind of phase transition, a liquid-liquid transition in glycerol. He shows you the data, a rather cluttered plot of various physical properties as a function of temperature. You note a vertical red line drawn at a particular temperature which is marked at the temperature of the phase transition. It is difficult for you to see that there is any signature of the phase transition in any of the properties except one, labeled "hyperturbidity," where there is a sharp change in slope at the red line.

You have never heard of hyperturbidity so you ask him to show you how it is measured. You go to the lab, and he shows you a setup in which a beam of laser light is passed through the sample and impinges on a small screen. He tells you that the hyperturbidity is related to the size and intensity of the spot on the screen. He adds that as the temperature changes you can detect those changes with your eye. He does a run right in front of you, but you can see no significant change during the run. When you tell him this, he explains that it takes practice to see the rather subtle changes. It is a skill that took him five years to master. He does another temperature run, but you still see no differences. He asks you for an opinion of his discovery. What should you tell him?

Commentary: Expectations (2)

This case attempts to raise the issue of a discovery that is probably merely in the mind of the discoverer. There are a variety of famous historical incidents of this type, including N-Rays and polywater. These were supposed scientific breakthroughs that turned out to be either delusions or artifacts. In this case, Dr. Mavis seems to have defined a new physical quantity, the "hyperturbidity," which only he can measure. He believes very strongly in this new phase transition, which no one else has seen. This finding is also suspicious because glycerol is a common, well-studied substance. On the other hand, more than one major scientific breakthrough was initially dismissed as being a delusion.

This case raises a delicate matter. Here you have a respected colleague who clearly believes he has made a major discovery. You need to be careful not to threaten his dignity. But scientific discoveries need to be reproducible by procedures that can be mastered by any interested scientist. It is true that special and delicate skills may be necessary, but a scientific discovery cannot be an occult phenomenon in which only a select few can participate. In this case you need to think of a strategy that will allow you to discover whether Dr. Mavis's discovery is real science or not.

Even though his discovery seems suspect, you do need to be fair to Dr. Mavis. The measurement might be subtle, and his discovery might be real. On the other hand, you have a responsibility to the scientific community to make sure that time and energy are not wasted on a nondiscovery. The polywater incident is instructive in this context.

A further discussion of these issues can be found in Felix Franks's *Polywater* (1981), Irving Langmuir's "Pathological Science" (1989); and Mary Jo Nye's "N-Rays: An Episode in the History and Psychology of Science" (1980). Gratzer gives a broad historical survey of similar incidents (2000).

CASE: A VERY USEFUL REFERENCE

As graduation approaches, you begin looking into fellowships at various universities. Your graduate work has been commendable, but you feel that you need another publication to make your application stronger. The work you've been doing lately is excellent material for publication, but you are having difficulty pulling everything together.

While working on the first drafts of the article, you relied heavily on the writings of other researchers, hoping that you'd be able to find your own words later on. Your advisor has stressed the need for exceptional work—the data are good, but the write-up will make or break the presentation. The pressure has been hard to handle and you have struggled for words quite often in your revisions. The material you "borrowed" from one particular article, which appeared in an obscure Australian journal, has been especially difficult to replace. The article was a lucky find. It didn't appear in your computer searches; you just stumbled upon it in the library one day. After deciding that you are completely stumped, you ask your advisor to read over what you have written and offer some suggestions for improvement. She is very impressed with your write-up, pointing out the sections that are not your own as excellent work. She comments that at this time she believes everything is going well and you should continue as you have been.

Surprised that your advisor did not notice the difference between your work and the work of others, you begin to think that maybe you could get away with the draft as it presently is if you just are sure to document the Australian journal as a source. But you realize that you have used too much of the article in that journal to have it as a reference. Someone might actually read the reference and discover what you have done. If you just didn't cite it, your article would be well written and ready for publication in plenty of time. No one, not even your advisor, is aware that you have been using the Australian journal. It did not show up in your searches. What should you do?

Commentary: A Very Useful Reference

The issue in this case is plagiarism. As defined by *Webster's II New Riverside University Dictionary,* "plagiarism is to steal or use the ideas and writings of another as one's own; to take passages or ideas from and use them as one's own." Because some passages in your article have been copied from the Australian journal, you have clearly committed plagiarism. You must make a choice about how you will handle the situation. Consideration of an appropriate solution should begin with the influences that might affect such a decision: the pressure you feel to publish, the application process, your advisor's comments or lack of them, the drive to succeed, and any others that may occur to you.

Once the constraints are known, you should consider various courses of action. Do you talk to your advisor or someone else about the matter and let this person help you? Do you keep it to yourself? Might this, even if you are trying to find a way to avoid plagiarism, affect other people's opinion of you or your work if you mention it to them? Are there ways to use the material other than plagiarism? Can anyone help you with the write-up? You might be able to get help from a campus writing center or from another student who is a more experienced writer. If you decide to leave the manuscript as it is, what might be the consequences of such an action on your career if the plagiarism is ever discovered? If it is not discovered? What are the legal implications? Will your application be hurt if you do not finish this article before submitting the application? Will it be enough to indicate that the article is currently being written up?

The best choice is to find some way to rewrite the plagiarized materials, retaining the important ideas and citing the Australian article as a reference. But even this option involves deciding to what extent you can use other people's ideas or words, even if you cite them, without plagiarizing. If you take the time to do this, even if you can only indicate the article as submitted in your application, you will have a stronger paper, and your professional integrity will not be compromised.

Plagiarism is discussed in detail in M. C. LaFollette, *Stealing into Print* (1992).

CASE: SUMMER RESEARCH PROPOSAL DEADLINE

Jonah is an undergraduate chemistry student at a small college. His parents are not very wealthy and have to work hard to keep him in school. In return, Jonah is a diligent and successful student. Because of his academic success, one of the chemistry professors has recently asked him to apply for a grant to perform research during the coming summer. One of the requirements of the agency is that the grant proposal actually be written by the student who will do the research, not by the faculty mentor. Not only will this be a great experience, but Jonah's father has been laid off from his job, which means that Jonah has to earn extra money to pay for school. With all the pressures of maintaining a high GPA and trying to conserve money, Jonah has not had the time to do much of the library research required to write the research grant proposal.

One day, as the deadline for the application approaches, you meet Jonah in the library frantically searching for information. He tells you that he has found something he thinks will work, but he does not understand it very well. Because he has his job and other homework to deal with, he thinks he does not have the time to read and understand the articles well enough to put things into his own words. He tells you that he thinks he will just copy some sections from the articles he has found directly into the proposal. Since it is unlikely that those reading the proposal will have read the articles, no one will notice and he will be able to get the proposal done before the deadline. What should you say to Jonah?

Commentary: Summer Research Proposal Deadline

The ethical question posed by this case study is whether Jonah's extenuating circumstances justify his actions. It is generally accepted that plagiarism is wrong because it is deceptive and selfish. Plagiarists steal another author's words or ideas and give the reader a false impression of their own expertise. When plagiarism occurs in a scientific forum, it also violates the integrity of the scientific community. It is true that Jonah's use of plagiarism to expedite the writing of his application will help him receive the financial aid he so desperately needs and provide him with valuable experience. But if Jonah's plagiarism is discovered, his proposal will probably not be funded. And if the funding agency reports the plagiarism to the college or the chemistry department, Jonah might face charges of academic dishonesty. At the very least, Jonah will get a reputation for dishonesty, which could hurt him when he applies for graduate school or for jobs. On the other hand, if the plagiarism is not discovered and Jonah is awarded the grant, he will have gained an unfair advantage over the other applicants who played by the rules.

There are many excuses for Jonah's plagiarism. He is normally an honest, hard-working student, but he is under enormous psychological pressure and is faced with a tight deadline. Jonah is a good person who is in a bad situation, not of his own making, and he deserves a break. For example, there is general agreement that it is wrong to steal, but we sometimes make exceptions. For example, we might say that it is acceptable for a person to steal food to feed a starving family. Although not as essential as food, education is very important, and without money to pay for it, Jonah could be deprived of an education. Plagiarizing a few passages from an article seems a minor transgression, fully justified by the unusual circumstances.

However, intellectual integrity is one of the core ethical principles of science. The process of science from the writing of proposals to the publication of results is based on trust. From the perspective of the scientific enterprise, Jonah's plagiarism is not a minor transgression. Since Jonah plagiarized a part of his proposal, he may not understand it very well. Consequently, he might not be qualified to perform the research. His scientific background or his abilities might not be adequate. As a result, both the professor and the granting agency will be cheated. Had Jonah been honest, the grant might have been awarded to a more capable student. While Jonah's financial circumstances might be desperate, there are options, such as student loans, that the case does not mention. Jonah's situation does not warrant compromising both his personal integrity and the integrity of the scientific community in this way.

Real-life moral decisions often involve compromise. There are cases in which deeply held moral principles can be compromised because of exceptional circumstances. For example, one might lie to save a life. Intellectual honesty is so fundamental to the process of science, however, that it is difficult to think of a situation that would justify an action like plagiarism.

CASE: PUBLICATIONS LIST (1)

A member of your research group will finish her degree at the end of the summer and is applying for permanent jobs. She is preparing her resume and comes to you for some advice. She has been quite productive as a graduate student, but none of her research has yet appeared in journals. One article has been accepted and three more are in various stages of completion. When you look at her list of publications, you notice that the three manuscripts that are currently being written are all listed as "submitted" to various journals. She asks you for comments on her resume. What should you tell her?

CASE: PUBLICATIONS LIST (2)

You are finishing a major grant proposal for the National Science Foundation (NSF). The preliminary work on which the proposal is based has gone well, but you have not yet completed any manuscripts to submit for publication. Several are in various stages of development and should be submitted in the next several months, assuming all the experimental work goes smoothly. Therefore, whenever you refer to these incomplete manuscripts in the proposal and in your bibliography, you list them as "submitted." Is this an ethically acceptable way to proceed?

Commentary: Publications List

Misrepresenting the status of a publication may seem like a minor issue, but it probably pervades the scientific community. In the first case the student has misrepresented the status of three papers on a resume; in the second a similar thing is done with a grant proposal. It is easy to dismiss these acts as "white lies" causing no harm to anyone. One could argue that by the time anyone reads the resume or proposal the papers will be in the mail. It is possible, however, that a project might fall apart at the last minute and only two of the papers are ever submitted. Then the lie becomes more serious, particularly if a judgment is made based on the person having three, rather than two, manuscripts submitted to journals.

One can misrepresent the status of publications in various ways. An article in the review process can be listed as accepted or in press. A paper that you plan to begin writing next month can be listed as in preparation. These are variations on the same theme.

Misrepresenting the status of a publication raises the larger question of the importance of truthfulness in representing one's accomplishments. All resumes, grant proposals, and similar documents are written to make the best case. Some exaggeration is common and expected; lies are not. When does exaggeration become deceit?

CASE: RETRACTION

Several years ago you published a paper interpreting the NMR spectrum of a novel compound. Since that time you have been working with related compounds and have discovered that the earlier work was incorrect. You had missed an important clue, and the unusual spectral features resulted from a completely different phenomenon. You are discussing this discovery with some colleagues at lunch. One person suggests that you publish a retraction, arguing that you should not let errors persist in the literature. Another person tells you not to bother, saying, "There is plenty of incorrect science in the literature. One paper more or less won't matter." Whose suggestion should you take, or is there a third, and better, course of action?

Commentary: Retraction

The issue in this case is whether you have an obligation to publish a correction or retraction of an article that you have found to be incorrect. There is no allegation of fraud; it was a simple mistake that neither you nor the referees of the paper were able to detect. There are lots of papers like this in the literature. In thinking about the right course of action, here are some questions to ask:

1. Who will be harmed or inconvenienced if you do not publish a retraction or correction? What kinds of harm might be caused by the incorrect information?

2. If you do publish a retraction, how will people know? Errata appear in a much later issue of a journal and often are not seen. Is there a better way to inform the scientific community of your error?

3. Has your error spread? Has your incorrect interpretation been incorporated into review articles or textbooks? Do you have an obligation to inform the authors of those publications of your action?

4. Should you consult with anyone, such as the editor of the journal, before proceeding?

CASE: DUST IN THE LAB

You are an undergraduate research assistant for Dr. Smith, a professor of inorganic chemistry. This semester, you and your fellow researchers are investigating new uses of a lead-containing compound. Your work involves repeatedly crushing the compound into a very fine powder.

While doing some online research, you come across a website that advises you to avoid airborne lead dust because prolonged exposure can lead to numerous illnesses and even death. You verify this information on the websites of three respected government health agencies. Everyone in the lab, including Dr. Smith, works with lead for hours every day, and you have indeed noticed powder drifting through the air. Like all chemists, you wear safety eye-goggles in the lab, but take no other special precautions. You decide to discuss the situation with Dr. Smith.

The next day, you visit his office, alone. When you tell Dr. Smith what you have found, he rises from his chair and closes the door. Dr. Smith appears to be really upset as he explains to you that he is perfectly aware of the dangers of working with lead he has a Ph.D. in chemistry, for goodness sake. And he goes on to say that the time you and the group spend working with the lead is not long enough to produce dangerous effects. You mention that the websites you found online indicate otherwise, but he questions the reliability of your sources. He also points out that the lab is on a very tight budget, so changing procedures and purchasing new safety equipment will not be possible. Dr. Smith concludes by assuring you that the lab is perfectly safe, so there is no need for you to trouble anyone else with your misgivings. He looks at his watch and realizes that he has an important meeting and must go.

After Dr. Smith leaves, you continue to sit in his office, stunned. You had complete respect for Dr. Smith; he is a top professor at the university and a leading researcher in inorganic chemistry. You don't believe that he would expose his students to dangerous conditions, and you decide to let the situation pass for the time being and go to the lab to begin the day's work. You notice that one of your fellow researchers, Andrew, is absent. You ask another colleague, and he says that Andrew called in sick for the next few days because he is experiencing a terrible headache and flu-like symptoms. You instantly recall that the symptoms of lead poisoning, which you read online yesterday, are precisely the ones Andrew is experiencing. What should you do?

Commentary: Dust in the Lab

The first step in approaching any moral problem is to clarify the facts. In this case, the facts seem quite clear, but it is always good to verify them. To confirm the information you found on the websites, you might confer with a toxicologist and a physician as well as check appropriate print sources.

You have several options. The first, and the easiest, is to do nothing. However, because Andrew has become ill, this is probably not the wisest choice. You and your fellow workers could be in serious danger.

Second, you could inform your fellow undergraduate researchers of what you know. Even though you would be blatantly disobeying the wishes of Dr. Smith, this situation is serious enough to prompt such behavior. Dr. Smith is potentially endangering the lives of his workers; there is no research important enough to jeopardize human lives. Thus, this is probably the most advisable course of action. You have a basic right to look out for your own safety and that of your coworkers, regardless of what Dr. Smith says.

Then, you and your colleagues could approach another person in the department, the department chair or the safety officer, about the situation and get an official opinion. Even though you respect Dr. Smith and enjoy your work, Dr. Smith is behaving irresponsibly, and his behavior could prompt punishment from the university. If the research is halted, you and your fellow researchers could potentially lose your positions. However, it is far better to lose your job than it is to suffer from lead poisoning.

CASE: SAFETY (1): LABORATORY CLEAN UP

Late one afternoon you are cleaning up after a successful day in the laboratory. Things have gone well, and you are looking forward to a dinner engagement with a special friend. Suddenly you notice a small flask of contaminated toluene that you forgot to put into the organic waste container. Carrying the flask to the hood, you notice that all the waste bottles are full. You look around the lab but find no empty waste bottles. Of course there are plenty of bottles in the storeroom, but it is located on the ground floor; it would take fifteen minutes to get there and back. You promised to pick up your friend at 6:30. If you leave immediately, you will have just enough time to get home, clean up and change, and get to your friend's apartment. You think to yourself, "It's just a few milliliters of toluene. I'll just pour it down the sink. It won't hurt anything." Is this action acceptable? Do you have other options?

CASE: SAFETY (2): A CARELESS COWORKER

You have just begun research in synthetic chemistry with Professor Holmes. He has asked one of his advanced students, John Watson, to help you get started on your project. You are quite impressed with John. He is bright, knows a lot of chemistry, and has excellent lab skills. He has been very patient with you, demonstrating the complicated techniques that you need to master. There is something about John that troubles you, however. He never wears his safety goggles in the lab, even when working with dangerous substances. You have always been careful about safety procedures, and John's disregard of them makes you quite nervous; you are concerned both about his health and about the possible consequences to you of his cavalier attitude. The one time you asked him about his lack of precautions he shut you off by saying, "Don't worry, I know what I'm doing." What should you do?

Commentary: Safety

In these two cases we address issues of laboratory safety. In research laboratories the individual has the primary burden of responsibility for following safety procedures. Deciding which regulation to follow is, in part, an ethical question. When you are in a hurry, it is easy to wash a small amount of seemingly innocuous waste down the sink. It seems a small transgression of the strict federal regulations on the disposal of hazardous wastes. Small amounts of waste, however, can eventually add up to large problems.

Beginning scientists need to learn that safety and environmental rules are important. Pouring the wrong substance down the sink, not wearing safety glasses, and not knowing proper emergency procedures all can lead to major disasters. The one time you fail to wear your safety glasses may be the time the reaction explodes. Following safety rules makes good practical sense, but it is also part of the personal responsibility expected of a professional. Not only might your disregard of proper procedures endanger your own health, it can also threaten the safety of others. For example, if you do not learn the proper procedures to follow in the event of a laboratory fire, other people might be victims of your bungled attempts to extinguish the flames.

The second variation, "A Careless Coworker," adds the complication of interpersonal relationships. What responsibility does an individual in a lab have to make sure others are following the appropriate safety procedures? At first glance it might seem that John is only endangering himself, but the real situation is more complicated. John's neglect of basic lab safety can endanger others in the lab. It can also lead to more serious consequences if the university or research institute has a safety officer or committee that can sanction the lab. There is also the possibility of lawsuit if someone is injured in an accident.

CASE: AN ACCIDENTAL SPILL

The government strictly regulates the use, storage, and disposal of radioactive materials in research facilities. In particular, there are certain steps that must be followed should any accidents—leakages or other contamination—occur. All students who might use or be exposed to radioactive materials are expected to have thorough knowledge of these procedures and safety precautions. When you and Mark, a friend from college, entered the graduate program at the university, you both attended the necessary seminars on radioactive substances. Your professors stressed rather stringently the importance of avoiding accidents in the radioactive labs.

Mark's first semester in the graduate program was a shaky one. His grades were some of the lowest passing scores in the class and his lab experiences were full of clumsy mistakes. Nevertheless, his desire to succeed has kept him working hard. His professors have, however, warned him that if some improvement isn't seen, they might be forced to release him from the program. You know Mark's performance as an undergraduate was exceptional; concerned about his future in the program, you agree to help him along as you can. As a result, the two of you end up together on your first experiment involving radioactive materials.

While carrying out the necessary procedures for the experiment, Mark makes a critical error that leads to the spillage of a substantial amount of radioactive sample. Mark immediately begins the clean-up process as he had been instructed, and within minutes he is through. The only remaining step is to notify the lab manager so that a record can be made of the incident. Mark, because of his academic situation, decides that he is not going to tell anyone about the spillage and asks you, as a friend, to keep quiet, as well. What should you do?

Commentary: An Accidental Spill

The primary question here is whether to follow the established safety guidelines. This question can be applied across the board, from day-to-day tasks in the lab to major experiments and special projects involving hazardous materials. Guidelines are established to ensure the safety of experimenters and to protect the public, but the extent to which they are followed is decided by those who must use them. It is not unusual for laboratory workers to cut corners and not follow strict safety guidelines.

Here, you must decide where your obligations are—with your friend or with the guidelines. Issues that should be considered include reasons for guidelines; consequences of following or not following them; whether or not there is a legal obligation to follow such guidelines; and when, if ever, a friendship or other relationship (such as a private one in which there might be an agreement to keep quiet about breaks in safety codes) comes before science or safety. Is Mark, with so many mishaps in the lab and poor grades, cut out for the graduate program at this time? Is it possible, since his undergraduate work was so good, that some external factor is affecting Mark's performance now? If this incident is reported, could it help Mark in some way? If you report it and Mark does not, do you run the risk of losing other people's trust? After all, you went against Mark, your friend, so why wouldn't you go against them?

Both ethically and legally, the proper thing for Mark to do is to report the incident and suffer the consequences. Your best course of action is to try to persuade him that honesty is really in everyone's best interest. Clearly, there is a temptation to protect Mark by pretending that nothing happened, but this is dangerous. If someone does find out, the consequences will be worse for both of you. Even more importantly, it is not good professional behavior to get into the habit of not following the established safety guidelines. Not only does this practice set a poor example, but ignoring safety procedures could result in a serious accident.

CASE: SAFETY DATA

Your best graduate student, a young woman, is engaged in an important project in synthetic chemistry. The target molecule is one that a number of major research groups around the world have tried to make without success. You have come up with a very clever idea that is almost certain to work. Synthesizing this molecule will certainly make your reputation and guarantee a promotion. In the course of your background reading you discover that one of the intermediates in the proposed synthetic scheme is closely related to a substance suspected of causing ovarian cancer. You are convinced that this woman is the only member of your group with the experimental skills to carry out this synthesis. If she makes and handles the intermediate, there is some risk to her future health. What should you do?

Commentary: Safety Data

This case raises some issues that will not be obvious to everyone. The obvious issue is the potential risk to the student's health. The first question, of course, is whether the advisor should disclose the suspected health risk. The next question is, Who should decide whether the student should proceed with the project? The obvious answer is that the student should be allowed to choose because her health and her career are at stake. On the other hand, because there is an unequal power relationship between student and advisor, she may not feel that she is free to withdraw if she feels that the risk is too great. This problem may be further complicated by the gender dynamics.

On the other hand, if the advisor tries to make the decision and take the student off the project, "for her own good," she can counter by arguing that he is taking away an important career opportunity, a chance at a major research success, which will help her get a good job.

Another issue is gender discrimination. Would you act the same way if the student were male and the threat was eventual prostate cancer or some other uniquely male disease?

CASE: PEER REVIEW (1)

You are studying the physical chemistry of pigments that play a role in vision. The information gathered in this study could eventually lead to cures of some eye diseases. In this competitive research area, new and important results appear in every issue of the major journals. The editor of *Proceedings of the National Academy of Sciences* sends you an article for peer review. The article details the research of three authors from the Harvard Medical School. The results are quite similar to the ones that your research group has obtained, but the three authors have performed a crucial experiment that you have not yet completed. Knowing the results will make it easy for your group to complete the necessary experiment quickly. You could then send your paper to a different journal. If you wait a week or so and send back the Harvard paper unreviewed, no one will suspect that you ever saw it. What course of action should you take?

CASE: PEER REVIEW (2)

You are in the midst of writing a major paper on the kinetics and mechanism of an important organic reaction, a project that has required a year of intense labor by your best postdoctoral fellow. When the paper on this "hot" topic appears, it should improve your reputation considerably. In the morning mail you receive an article to review from the *Journal of the American Chemical Society*. Upon opening the envelope you learn, to horror, that a senior colleague at another university has written a paper on the same reaction. You quickly read the paper and discover that it looks as if you have been scooped. Because this is a novel system, you realize that you could raise objections to this paper that would delay its publication for months, giving you time to submit your own article to another journal. On the other hand, since the work is very similar to your own, you know that it is correct. What do you do?

Commentary: Peer Review

These two similar cases raise a fairly common ethical problem in research: What do you do when you have been "scooped"? In a competitive research area, several groups are likely to be working on similar or identical projects simultaneously, sometimes without knowing it. In Peer Review (1) there is the temptation to cheat, either by using results from the paper you have received to review or by delaying the paper to allow yours to be published first. In Peer Review (2) a colleague has written a paper similar to one you are working on and you have received it for review. If this paper comes out first, yours might not even be published. As in Peer Review (1), one possibility is to delay this paper so that yours will come out first.

There are a number of factors that might induce someone to choose this alternative. For example, a young faculty member who is under enormous pressure to publish because of an upcoming tenure decision might decide that it is worth the risk. A more senior researcher might be struggling to maintain productivity and federal funding. Usually, it is relatively easy to raise enough objections to delay the publication of a paper. Because referees are anonymous, the original authors are unlikely to know who wrote the critical comments. Such an action, however, clearly violates the ethical standards of the peer review process.

If this ethically questionable route is rejected, as it should be, the next question is that of fairness. There are two sides to the fairness issue. You must be fair to the authors of the other paper, but you also must be fair to yourself. You have done the work independently and deserve some credit for it. There are several options. One is to accept the fact that you have been scooped and go on to another project. Another is to contact the other authors and agree to publish back-to-back papers or a joint paper. They might not agree, of course. It may be that your colleague is very competitive and will not agree to any course of action that recognizes your work as equivalent. In addition, before you begin such negotiations, you need to talk to the editor of the journal because you will be breaking the confidentiality of the review process. If you do decide to try one of these courses of action, you can no longer be an objective reviewer of the paper, so you will need to return it to the journal unreviewed. Out of fairness to the journal and the author, you should do this as quickly as possible. A third possibility is to alter the course of your project to salvage a publishable paper that is sufficiently different from the one that you received to be acceptable to the journal. All three are ethically viable alternatives.

CASE: GRANT PROPOSAL

You are in the midst of a project to synthesize a new compound. The target molecule is one that many groups around the world have tried to make and failed. Your synthetic route is quite clever. Unfortunately, the graduate student working on the synthesis has not been able to get one of the steps to work well enough. She can only obtain a 5% yield, which is not enough to go on to the next step. Once this problem is solved, the rest of the steps are quite straightforward. You have suggested everything that you can think of, but nothing seems to work. In the midst of your frustration, the National Science Foundation (NSF) sends you a grant proposal to review. As you read it, you realize that the author has the solution to your student's problem in the proposal. A novel three-component mixed solvent system that he is using in a different context should increase the yield to at least 50%, which will be good enough. You are tempted to run into the lab and tell your student to try this solvent system immediately. Then you remember that grant proposals are confidential information. What should you do?

Commentary: Grant Proposal

The major issue is expressed in the penultimate sentence: What are the limits on your use of confidential information? Reviewers of grant proposals and papers are reminded that the documents are confidential and that the information they contain should not be used. Having read the proposal and realized that the solution to your problem is in it, you cannot forget what you read. You have to be fair to the author of the proposal, but you also have to be fair to yourself and your student. Your project is quite unrelated to the one in the proposal, so you are not infringing on anyone's research program. If the proposal is funded, the writer will be able to carry out the work with no fear that you will have done it first. On the other hand, you have not seen that solvent system described anywhere else, so it will be fairly obvious where you learned about it.

One possibility is to ask the author of the proposal for permission to use the system. If you do that, however, you compromise the confidentiality of the peer review process, and the NSF program director might not be happy with you; thus your own future funding may be jeopardized.

Another possibility is to ask the NSF program director to act as an intermediary and request permission to use the solvent system in your work. In this way the confidentiality of the review process can be maintained. Once your work is published, the author will learn your identity, but by then, the final decision on the proposal will have been made.

Active scientists see confidential information often in the articles and proposals they review. Since it is difficult for them to forget what they have read, these confidential documents have an influence in their future work. The difficult ethical question is to distinguish between the influence on thought and theft of intellectual property. Dorothy Nelkin's book *Science As Intellectual Property* (1984) and Corrine McSherry's more recent *Who Owns Academic Work* (2001) explore intellectual property issues in depth.

CASE: COLLABORATION

A postdoc, Patricia Jones, has been working in your lab for only two months but has already proven herself as a productive, respectable researcher. At present, she is working on a project that you initiated which, once completed, should strengthen your reputation. Dr. Sanders, a friend and colleague from the university Patricia attended, has shown much interest in the work you and Patricia have been doing. He proposes a collaboration between your lab and his. It will most likely add a new dimension to the study and will increase the funding available for necessary experimentation. Without much hesitation, you agree to Dr. Sanders's proposal.

When you tell Patricia about the good news, she is not too pleased. A close friend and advisor of hers from graduate school worked with Dr. Sanders on two occasions. Because of several things that happened, she is convinced that Dr. Sanders is the type of person who will steal good ideas and make them his own before anything can be done about it. You assure Patricia that she must be mistaken, or that her advisor was, but she holds to her beliefs and strongly urges you to reconsider. You are surprised at the intensity of Patricia's reaction. Dr. Sanders is someone whom you know and respect, but in the short time Patricia has worked with you, she has shown good sense and mature judgment. Perhaps your assessment of Dr. Sanders is wrong. What should you do?

Commentary: Collaboration

How much attention should you pay to the comments of this postdoc who has only been in your lab for two months? Patricia made strong allegations about Dr. Sanders which seem inconsistent with what you know about him. Her comments are based on things she learned from her former advisor. Perhaps that person is wrong or has some private quarrel with Dr. Sanders and is trying to ruin him. On the other hand, the allegations could be correct, and your good opinion of Dr. Sanders may result from not knowing him well enough.

The obvious first step is to get more information about Dr. Sanders and to do it discreetly. If Patricia's allegations become widely known, they could harm Sanders's scientific reputation. But to get the information you need to evaluate her claims, you need to talk to both Patricia and her former advisor as well as to others in the community who might know more about Sanders. You will have to ask some potentially embarrassing questions.

No matter what you find, the situation is delicate. If Patricia is right, it is in your best interest to break off the collaboration with Sanders, but you will have to give him a reason for changing your mind. Do you tell him the truth, which will probably lead to denials or excuses and perhaps an angry exchange, or do you make up a more benign excuse to avoid any conflict? After ending the collaboration, should you take any further steps? These are difficult questions. Although Sanders's behavior is certainly a violation of professional etiquette, if not ethics, there is only a small probability that you can have any effect on him, particularly at this stage of his career. On the other hand, it would be good to try to keep others from being victimized by his actions.

If Patricia is wrong, you have a responsibility as her postdoctoral mentor to correct her error and help her understand that spreading unsubstantiated rumors is inappropriate professional behavior. This is an *ethics moment*, an opportunity to discuss a question of professional ethics. But the discussion should be handled delicately. It appears that Patricia has been influenced by her former advisor to whom she will probably have a strong loyalty. You don't want to destroy that relationship. Among other things, Patricia will need to call on the advisor for recommendations for future positions.

This is an example of an ethical problem that does not seem to have a clean solution, largely because of the personal relationships that are involved.

CASE: AN OVERHEARD CONVERSATION

You are a new graduate student in biochemistry, and you have been working on a project for weeks without making very much progress. Despite the fact that you have gone over your methods again, you cannot figure out what you are doing wrong. You discuss the matter with Dr. Waters, your research advisor, who reminds you that in graduate school you must learn to find the answers for yourself. However, Dr. Waters does, however, agree to observe you during the next experiment.

You follow the outlined methods and, aware that you are being observed, take extra precautions to make certain that the experiment goes the way it should. Nevertheless, the results still do not compare with the published data. Dr. Waters does notice a problem with your execution of the procedure, but he wants to give you a bit more time to figure it out on your own. After running the experiment again and becoming frustrated with the results, you decide to go to the graduate lounge to rest for a bit. When you get there, you hear two other graduate students talking.

By the time you are just outside the door, you realize the students are talking about the experiment you are working on, and you decide to listen for a bit. Eventually you hear what you have been doing wrong. One of the students, Jon, who is also working with Dr. Waters, but with whom you have had several disagreements the past few days, also mentions an idea he has for a related project. If you were to stay in the lab all night, you could probably complete your original project and get a fair portion of Jon's project done in time to mention it to Dr. Waters first. You know that Jon has not yet mentioned the idea to Dr. Waters, and you could just say you had a very productive evening in the lab. You are sure that no one has seen you listening and doing this would give you a chance to get ahead. What should you do?

Commentary: An Overheard Conversation

This case raises the issue of using other people's ideas for your own benefit. Several variations of this theme are seen in other cases: plagiarism, borrowing ideas from grant applications or reviewed articles, and using information received from a faculty candidate. Here you are contemplating whether you should use something that you overheard that will allow you to complete a project you have been struggling with. You have also heard an idea for a related project. You must decide how much, if any, of the information you will use in the lab. You do not have to consider breaking confidentiality rules or legal guidelines for plagiarism.

There are actually two questions. One of the things you heard was the solution to your current problem with the procedure. Dr. Waters could easily have helped you earlier that day, but he made a pedagogical decision to give you time to figure it out on your own. Now that you know the answer, there is no reason for you not to use it. It was hardly a secret. Although it might have been more beneficial for you to have discovered it yourself, that opportunity is now lost.

The temptation to use the idea for the related project is a different question. If you begin to work on Jon's idea without his permission, you are stealing his intellectual property and violating scientific etiquette, if not ethics. Of course, there will be a real temptation to begin work on this idea. It is a natural extension, so no one will be able to prove that you didn't come up with the idea independently. If you are able to make it work and obtain interesting results, Dr. Waters and the other members of the research group will be impressed. The project could become an important part of your Ph.D. work and help launch a successful career.

On the other side, your success would be tainted, not the result of your own creativity. In addition, competition can create tension in a research group. Even if both you and Jon had come up with the idea independently, it would be better if the decision as to who should pursue the project be made openly in consultation with Dr. Waters. The smooth functioning of the group requires trust. If Jon begins to suspect that you have stolen his idea, the interpersonal relationships in the group could become tense, making it difficult for all of you to pursue your research.

While no one "owns" nature, there are standards of etiquette in the scientific community, and stealing ideas is behavior that is frowned upon. Scientists who gain a reputation for such behavior are not well respected. For that reason alone it is best to be very careful in using information you have overheard.

CASE: THE HELPFUL CANDIDATE

The hiring freeze at your university was recently lifted, much to the relief of your department, which needs to fill a long-vacant faculty position. The application response has been overwhelming, but the field has been narrowed to five candidates who will be interviewed in the coming weeks. Several of the candidates have backgrounds in research closely related to that of some of the present faculty members, including yourself. You are eager to see what work they have done and ideas they could bring to the department.

The third candidate to be seen is Dr. David Stull, from Seattle, Washington. His work is very similar to yours, and during the course of interviews and tours, he reveals information that is of tremendous interest to you. He has the necessary credentials for the position, and his presence could mean an increase of funds in your field. You announce your support for Dr. Stull in a departmental meeting, commenting that his research and ideas, combined with yours, could bring much recognition to the university.

In the following weeks, the remaining candidates are interviewed. You haven't had much time to look into the statements Dr. Stull made that could benefit your research, but you are looking forward to doing so in the near future, maybe when Dr. Stull arrives in the fall. When the final decisions are made, however, Dr. Stull is not appointed to the faculty position. What should you do about following up on the ideas and information you received from Dr. Stull?

Commentary: The Helpful Candidate

The primary issue is given in the question, Should you use the information and ideas that Dr. Stull provided now that he will not be a faculty member at your institution? This case is similar to the two variations of Peer Review in that it involves using other people's ideas, but here peer review confidentiality is not a problem. Dr. Stull only casually mentioned the idea, not in a detailed discussion or even in an interview. He was just touring the facilities. You must decide what your next step should be. Do you contact Dr. Stull despite the fact he wasn't hired and discuss the ideas further, asking how he feels about you using them in your own work? Do you suggest working together on a project? How would that approach affect your department? Dr. Stull? Or do you investigate on your own and use the ideas without any recognition of Dr. Stull? If so, how would you justify such an action?

The delicate point is that Dr. Stull was not offered the position in your department so contacting him to discuss this matter might be awkward. On the other hand, a collaboration might be beneficial to both of you. Using his ideas without talking with him first is a form of "poaching," perhaps not a strict violation of professional ethics but certainly a breach of etiquette.

CASE: AN OLD PROBLEM

Sam worked for Acme Chemicals for twenty years but decided to switch jobs to a new chemical company for better pay and a nicer location. Soon after he started his new job, he found himself working on a project similar to one he had been involved with at Acme. After a few weeks, the project came to a halt when his team could not solve an intricate problem that his former team at Acme had solved rather simply and cheaply. Sam didn't want to steal the idea, but management was putting increasing pressure on his team to solve this puzzle, and with the recent talk of downsizing, Sam's team was growing desperate. Is it acceptable for Sam to show his team the solution to the problem that he had helped develop at Acme? If not, can he "guide" his team in that direction and let someone else in the team find the solution he already knows?

Commentary: An Old Problem

Ordinarily, employees are required to sign confidentiality and patent agreements when they join a company. These agreements prohibit them from revealing proprietary information and company secrets both while working for the company and after they leave. If Sam has signed such an agreement, saying anything might violate the terms of the agreement and make him subject to legal action. Acme could sue him. Therefore he should certainly review the relevant documents to determine the nature of the restrictions.

Even if Sam has not signed such an agreement or if the information that Sam might provide is not so restricted, there is still a moral question. Is the information confidential or proprietary, or is it just public knowledge that the team at Sam's new company could find if it only knew where to look? These are questions that must be decided case by case. If the process is protected by a patent or by a clear confidentiality agreement, then the answer is clear: Sam should not say anything. If, on the other hand, the solution is a simple application of a well-known scientific or engineering principle, then Sam can tell the team at his new company anything they need to know. The situation in this case seems to be in the vast gray area in between, so a detailed analysis of the facts is crucial.

The additional complication is the economic pressure. Sam's new company would really benefit from solving this problem. If they fail, there is the chance that members of the team, including Sam, could lose their jobs in a downsizing. This fear could push Sam to do something illegal or unethical. This is a kind of conflict of interest, in which Sam's self-interest (keeping his job) or the company's self-interest (a successful new product) conflicts with standards of professional ethics.

These questions are discussed in the engineering ethics literature, for example by Davis (1998) and Harris, Pritchard, and Rabins (1996). For a provocative opinion, see the editorial by Roald Hoffmann (1997).

CASE: REVIEW ARTICLE (1)

You are a well-known analytical chemist specializing in chromatography. During your career you have developed a number of new instruments, which the Chromo company has eventually marketed. Chromo has generously supported your research for many years and also retains you as a consultant, providing a nice supplement to your university salary. You have just received a letter from the editor of a journal that publishes critical reviews in analytical chemistry. The editor has asked you to write a review on current chromatography techniques. In this review you will be expected to compare the various commercial instruments and to make judgments about their capabilities. Here are questions to consider:

1. In view of your financial involvement with the Chromo company, should you agree to write this review?

2. If you do write the review, should you mention your relationship to Chromo to the editor? Should you also disclose your relationship in print?

CASE: REVIEW ARTICLE (2)

After ten years with Chromo, a major manufacturer of chromatography equipment, you and the management decide that you should return to school to obtain the Ph.D. You decide to attend a major midwestern university and do your graduate research with Dr. Collum, a leading figure in chromatography research. Chromo promises you a job after you earn your degree.

Shortly after you begin your research with Dr. Collum, she is asked to write a critical review of current chromatography techniques for a major journal. In this review you will be expected to discuss commercial instruments, including those made by Chromo, and make judgments about them. Because of your extensive background, Dr. Collum asks you to help with the review. Eventually you will be listed as a coauthor on the published review. Consider the following questions:

1. Because of your relationship with Chromo, should you decline the opportunity to participate?

2. If you do participate and become a coauthor, should your relationship with Chromo be mentioned in the article?

Commentary: Review Article

The issue here is conflict of interest. Scientists are expected by their colleagues and society to render impartial professional judgments based on a critical review of evidence. This is what is being asked of the author of the review article. Conflict of interest results from some tug on the scientist (or other professional) that makes his or her professional judgment less reliable. In this case your financial relationship with Chromo is the obvious tug on your judgment. Put another way, there is a secondary interest, your financial relationship with Chromo, which potentially conflicts with your primary interest as an objective reviewer. Although you may think that you are being completely objective, subtle influences may come into play.

On the other hand, if you decline to write the article, the editor will ask someone else. Will that person be equally qualified? If the other possible authors are less qualified than you, are you being fair to the scientific community in not writing the article? Will that person have some relationship to another company, one of Chromo's competitors? Are you then being fair to Chromo by declining the opportunity to write the review?

There are three ways to handle conflict of interest. First, you can avoid them. In this case, that means not writing the article. The second is to divest yourself of the external influence. In this case, you could sever all financial ties with Chromo. Unfortunately, the long-term relationship with this company might still influence your judgment, one way or the other. The third method is to reveal the possible influences by stating your relationship to the company in the review article. This disclosure at least gives the reader a fair warning of the possible influences on objective judgment.

The second case presents the same issues, but from the perspective of a graduate student. Since the graduate student will be the junior author, the consequences of the conflict of interest may be less serious, but still the problems remain since the student has an influence on the content of the article.

No one course of action is best. In this case, each of the three possibilities can be defended. The best course of action for the individual will depend on details not specified in the case statement. For further information on this issue, you may want to read Michael Davis's "Conflict of Interest" (1982) or Dennis F. Thompson's "Understanding Financial Conflicts of Interest" (1993).

CASE: REVIEWING AN ARTICLE

You are a prominent university researcher in biochemistry, working on antisense therapy for skin diseases. Your work has resulted in several patents, and you and a colleague enlisted a venture capitalist and have started a company to commercialize your invention. The company has gone public, but you still own 10% of the stock. You also retain your university position as professor of chemistry.

One day the editor of a major journal in the field sends you a manuscript to review. The two scientists who wrote the article are employed by a company working in the same area as yours. In fact, the two companies will compete in the same market if they are successful in bringing out products. The paper reports interesting advances in the biochemistry of the molecules that these companies are trying to commercialize.

Should you agree to review the paper? If you do, should you reveal your financial stake in your company to the editor of the journal?

Commentary: Reviewing an Article

This case raises an issue that is becoming important, particularly in biotechnology, but also in chemistry. More and more university professors are directly involved in commercializing their ideas by forming venture companies. Often the researcher holds significant equity in the company and therefore has an enormous financial interest in its success. The question in this case is whether that financial interest gets in the way of objective scientific judgment.

Scientists from a rival company have written a paper that you have been asked to review. Can you judge it fairly? At first glance you might think that the commercial ties are irrelevant because this is a journal publication and not a product. On the other hand, a major publication from a rival company can increase the value of their stock or aid in their pursuit of new investors. Such considerations might influence your review.

If you do decide that you should do the review, should you inform the editor of your financial relationship with your company? Does the editor need this information to put your review in context? Should journals adopt a policy requiring reviewers to disclose their potential conflicts of interest? This question is discussed in some detail in a special report on "Conflicts of Interest," published in the July 31, 1992, issue of *Science* (Koshland 1992; Barinaga 1992; Marshall 1992)

Conflicts of interest often involve financial considerations. For example, consider the following situation. Company A is about to make a public stock offering. Scientists from that company have submitted a paper reporting a major advance to a leading journal. The paper is sent to reviewer B who happens to own a large block of private stock in that company. The editor is unaware of that stock holding. Reviewer B writes a glowing review. The article is published, and one week later, when the stock is offered for public sale, reviewer B sells his shares at a tidy profit. Stock analysts suggest that the paper had a positive effect on the value of the stock.

Not all conflicts of interest will involve such an immediate benefit, but the entangling of private financial gain with scientific judgment is an increasingly important ethical issue.

Some of these issues have been discussed in my "Gifts and Commodities in Chemistry" (2001) and by Brian P. Coppola in "The Technology Transfer Dilemma: Preserving Morally Responsible Education in a Utilitarian Entrepreneurial Academic Culture" (2001), and Michael Davis in "University Research and the Wages of Commerce" (1999).

CASE: COAUTHOR (1)

You are preparing a manuscript for publication. The theoretical treatment of the subject of the paper is outside your competency, so you have discussed much of its content with a senior colleague who is a theoretician in this field at another university. All the experimental work, however, has been done in your laboratory. As a newcomer you have brought a unique perspective and a new set of experimental methods to bear on the problem. You are confident of the experiments and the conclusions you have drawn but are afraid that the reviewers in this new research area might be skeptical of your work because it is unusual. Therefore, to strengthen the paper, you add the name of the senior colleague to the list of authors. His name should lend additional credibility to the paper. Because of your extensive conversations with him about the paper, you assume that he will be happy to be included as a coauthor, so you don't bother to ask his permission before sending the manuscript off for review. Is this acceptable?

In thinking about this case, here are some questions to consider:

1.   What advantages might be gained by adding this name to the paper? Are any of these advantages unfair?

2.   What responsibilities are placed on the senior colleague if you add his name to the paper?

3.   What are the possible negative consequences of this action for you and for your senior colleague?

Commentary: Coauthor (1)

This case raises the issue of honorary authorship. This practice is more common in the biomedical sciences than in chemistry, but it certainly occurs in all fields. In this case the primary author is a newcomer to a field and hopes that the paper will gain more status if the name of a senior colleague appears on it. The powerful colleague's name might help the paper through the peer review process. Once the paper is in print, it might be more widely read. Scientists scanning the journals tend to key on names they know, and reviewers are more inclined to accept controversial results from established figures than from newcomers. These are several advantages that might accrue from this seemingly minor deceit.

It is also important to consider this action from the point of view of the senior colleague whose name has been added to the paper. It may seem like a gift. He gets another publication for his bibliography for relatively little effort. On the other hand, being a coauthor usually implies that you have had a major role in the paper, understand it, and agree with its conclusions. Your colleague may be asked to comment on or defend a paper he has not even read. He may not agree with everything you said. If there are major or minor errors, readers will assign part of the blame to him. Because he is the major figure, he may take more criticism than you.

The tactic may backfire, however. If the paper is a major success, it may increase his reputation much more than yours. You and any students who worked on the paper might be looked on as minor contributors. If you then try to obtain funding or publish papers without your colleague, your work may not be as well reviewed because it is assumed that you don't have the expertise to succeed in this new field.

Many of the issues related to honorary authorship are explored in "The Noncontributing Author: An Issue of Credit and Responsibility" by Roger P. Croll (1984). Another resource is John Hardwig's "Epistemic Dependence" (1985).

CASE: COAUTHORS (2)

You are working as a postdoctoral fellow in the research group of a famous scientist. She has asked you to work on a particular problem of interest to her. During the course of your work on this problem, you discover a small theoretical issue that you can resolve quite easily. You perform the calculation and write up a short note to send to a journal. Because you have done this completely independently, you do not include your advisor as a coauthor. When you show her the note, she praises the work, but demands that she be included as a coauthor on the paper. Is this right?

Commentary: Coauthor (2)

This case also explores the honorary author issue, but from a slightly different perspective. Here the research advisor demands that her name be on the paper even though she did not contribute to its content. The usual justification is that the advisor provided the financial support and other resources so that the work could be completed. The moral question, of course, is whether this is a sufficient condition for a coauthorship. I expect that a large number of senior scientists would immediately answer in the affirmative.

As is discussed in the commentary to Coauthor (1), having her name on the paper can be either an advantage or disadvantage for the advisor. The same is true for the student.

In addition, because of the unequal power relationship between student and advisor, the student may not be in a position to say no to the demand. A research advisor has enormous power over the student's future. A gender difference may also have an effect on the dynamics of the power relationship. A female student might not feel that she can confront a male research advisor.

A related question is the order of the authors. What is the appropriate way to list the authors on a paper? To some extent, the traditions of a particular field determine the answer to this question. In some disciplines the senior or most important author is first; in others, the senior person is last. In citations, however, the first author is usually the one who is remembered, so that position takes on long-term significance. Although on the tradition of the field can provide guidance, the crucial issue is fairness, giving credit where credit is due.

CASE: UNDERGRADUATE COAUTHOR

Chris, an undergraduate researcher, is near the end of the experimental phase of a kinetics project that Dr. Cain, your graduate advisor, has suggested. Chris's data, if sufficient, will be used in a paper detailing your current research. Not only because his work is relevant, but also because it will enhance Chris's applications to graduate schools, Dr. Cain has mentioned including Chris as a coauthor in your upcoming paper. He has made it clear that he expects you to support Chris and be available for assistance so that the authorship will be for a valid contribution.

In an attempt to do this, you have on several occasions asked Chris about his work and suggested ways to improve his techniques. Chris, although friendly, does not seem to be too receptive to your suggestions and has not been willing to discuss the results with you.

After Chris has left for the day, you notice his notebook and computer printouts on the workbench. Curious about the results of the experiments, because of their importance to your work, you begin to look through Chris's papers. The results do not make any sense and do not suggest any of the necessary correlations for your paper. Out of frustration and the need for good results, you decide to stay late and repeat some of Chris's experiments as they are outlined in his notebook. The results you obtain are different from Chris's and provide the necessary correlations.

In the morning, you inform Chris of your findings, making every effort to be supportive rather than accusatory, because you are well aware that his difficulty with the experiment is most likely due to a simple technical error. Chris, however, is offended by your snooping and asks you to mind your own business. You try to explain that the results of the experiments are your business, but Chris is not responsive. He simply continues his work. When you mention the situation to Dr. Cain, he tells you to take it easy with Chris because he is the son of a very good friend for whom Dr. Cain is doing a favor. Dr. Cain tells you just to repeat everything that Chris does and use that data in the paper. Chris will have done work on the project so he can still get the authorship as discussed.

You are appalled by Dr Cain's comments and reasons for having Chris in his lab. And you certainly do not want to give credit to Chris for work not properly done. What should you do?

Commentary: Undergraduate Coauthor

The issue in this case is authorship. Unlike other cases in this book, however, the authorship is not for a prominent researcher or advisor, but for an undergraduate who happens to be the son of the advisor's friend, for whom he is doing a favor. It is clear that the authorship is a gift the advisor wishes to give, not a recognition of any significant contribution to the paper, so the ethical questions are a bit different from cases in which judgments have to be made about the relative importance of contributions to the research. An important issue is how listing Chris as an author affects the paper, the graduate student, the advisor, and Chris. Because he is an undergraduate, it is unlikely that Chris would be held responsible for any problems that might occur or receive much credit for the work done. Nonetheless, the authorship will certainly look good on an application to graduate school. Might there be negative consequences if Chris proves to be a poor research student in graduate school?

The primary ethical and practical issue here is how one should handle the advisor's reaction. You feel that the professor is wrong, but if you challenge him, your relationship may suffer. What might be the best way to approach him? Because it is your paper that is being discussed, you have to consider your own interests in the situation. Can you safely confront your professor again? Should you go above him and talk to the department chair? Might it be possible to talk to Chris and persuade him to decline the offer for an authorship? These and other options should be considered before deciding on a course of action. On the one hand, adding Chris's name to the paper is a minor irritation. The scientific community will recognize that the work was primarily yours. On the other hand, Chris will get not only credit for something he has not done but unjustified recognition as an aspiring scientist.

Because your professional future, at least in the short term, depends on your research advisor, you need to be careful. Is the principle of appropriate credit sufficiently important to risk undermining your relationship with your graduate advisor?

CASE: PRESS CONFERENCE

You and your research group have been engaged in work on the development of new high-temperature superconductors. In this very competitive, fast-moving research area new results seem to appear every week. Your group has made a new class of materials that have the possibility of being superconducting at temperatures near that of liquid nitrogen. You have kept the work quite secret and have just finished writing a communication to a major journal; it can be sent out within a day or two.

Excited by your discovery, you report it to your department chair. She is equally excited and phones the university research office to alert them. They suggest that you hold a press conference the next day to report your results to the general public. Should you agree to this proposal?

Commentary: Press Conference

"Publication by press conference" has recently become an important issue. In fast-moving fields where the results have news value, researchers sometimes publicize their accomplishments in the popular press before their papers have been subjected to the usual process of peer review. The cold fusion incident is perhaps the most famous recent case. This practice is generally discouraged. Some journals will not publish an article if its contents have been previously released to the press.

On the other hand, some scientific results are of interest to the general public. People are quite interested in the latest developments in medicine. A number of public policy issues involve scientific questions. Tax dollars fund much U.S. research, making it a legitimate object of public scrutiny. Scientists do have a legitimate self-interest in ensuring general public support.

This case raises questions at a variety of levels. Some of them are

1. Is it appropriate for the department chair to call the university research office without your permission? What are her (or any other colleague's) obligations to keep your results secret? Is it your responsibility to ask for secrecy?

2. Under what circumstances is it appropriate to release scientific results to the press? At what stage in the process is it legitimate—after the paper is written, after peer review, or after publication?

3. What are the responsibilities of reporters, editors, and scientists to make sure that the information in the press is both accurate and understandable?

4. What are the legitimate interests of the general public in the broad dissemination of scientific results?

5. Should the funding source have any influence on your decision? A granting agency might have regulations concerning this matter. If the funding came from private industry, the company's interests in the research might be compromised or enhanced by the public announcement.

Many of these issues are explored in *Selling Science* by Dorothy Nelkin (1987).

CASE: PRESS RELEASE

For the past year, you have been working on an intense study to find a cure for Ekoms disease, a condition that has infected an entire island population and was discovered only three years ago. Your partners in the research, Dr. Palmer and another member of his lab, have slowly lost interest in the project as new grants have been funded, but occasionally stop by to see what progress has been made. This declining participation has been difficult for you to accept, but you have not made any effort to discuss how you feel about it. Regardless of their present ventures, they too have invested much time and money into the Ekoms study and are eager to see it finished. The public, as well, is awaiting your results, hoping that a cure for Ekoms will soon be found.

You know that you are close to finding a cure, but the data are not yet conclusive. While you are working on some new variations of the product you've developed, Dr. Palmer stops by. He looks over the latest information, comments on how close the two of you are to fame, and then leaves. A week later, while analyzing the results of your experiments, you decide that, at last, you have it—a cure for Ekoms. You need to gather more experimental evidence to provide proper verification, but you know that this time it has to be right.

Despite your excitement, you do not tell Dr. Palmer about the findings immediately because you want to enjoy your pride without having to share it for a while. Instead, you decide to celebrate this evening with a close friend who also happens to be a journalist for the city's top newspaper. While discussing how you have done most of the experimental work but will not get most of the credit because of Dr. Palmer's involvement, your friend comments that it would be great if you just bypassed the normal publication routes and let her write up your discovery in a story to be published in a matter of days. She could conveniently leave Dr. Palmer's name out of it. Should you agree to this?

Commentary: Press Release

Several issues are raised in this case, but the primary one is whether to bypass all other persons and the normal publication routes involved in the release of new scientific breakthroughs or evidence. If Dr. Palmer were still as actively involved as he was in the beginning of the collaboration, you might not feel as you do—angry at his lack of effort—and you might not even be tempted by an opportunity to take all of the credit for the discovery as your friend has suggested. But could letting this news get released by untraditional means cause more difficulty than it's worth? Might you be scrutinized and suffer more than if you lost some credit to Dr. Palmer? Or could this push you into the limelight enough to make it worthwhile? You have not actually finished all the necessary verifications of the product, but if it all goes well, you'll definitely have fame-making material in your hands.

But what happens if the verifications do not actually come through? How close to completing experimental work do you have to be before communication to the scientific community can begin? Communication to the public? What types of reactions would you get from each group with an early release of important, groundbreaking information? What will be the consequences of retracting such claims at a later date? How might a retraction affect the public view of scientific research and funding? How might it affect your colleagues?

When you collaborate on a project with your colleagues, how much does your personal relationship play into the professional one? Is it right to collaborate, leave most of the work to another person, but still take equal credit for the result? What responsibility do you have to maintain the collaboration after the other members have faded out of the picture a bit? Should you discuss these issues with your colleagues? If so, how might you go about it? If not, why wouldn't you discuss it?

Another consideration, of course, is peer review. The norms of science demand that all articles be scrutinized in the peer review process before being published. It is possible that you have missed something or made a mistake which the reviewers or the editor might catch. Is it responsible to release scientific findings, particularly results that might have an impact on public health and welfare, before they are validated by the appropriate scientific community? Although there are temptations to bypass them, the time-honored process is the responsible way to proceed.

CASE: GRANT APPLICATION

While doing preliminary research for a grant application, you make several discoveries that you would like to investigate further. The issues are controversial, however, and it is not likely that the institution you are applying to will fund these projects. On the grant application, you decide to limit the information you provide to the reviewers, leaving out a portion of the costly but necessary preliminary work that has already been completed. You have collected more than enough evidence in support of your hypothesis to be funded. You plan to propose the already eliminated preliminary work as some of projects to be completed under the new grant. The funds granted for the work already completed can then be used for the controversial investigations. Is anything wrong with this decision?

Commentary: Grant Application

The major issue here is truthfulness in a grant application. You have two things in mind. First, you are proposing to do work that is already done, which obviously has some advantages. You know that part of the project will succeed; you will be able to publish it rather quickly and impress the granting agency. Technically, however, a grant proposal is supposed to ask for funds to pursue work that has not yet been done, so your proposal is disingenuous, at best.

Second, you actually plan to use the money to pursue a controversial project that you fear will not be supported if you ask for funding explicitly. Because granting agencies expect you to work on the project you actually proposed, this is also dishonest. You might argue that the peer review process forces you into such deceit. Granting agencies tend not to fund risky work, and they want to see results quickly and in quantity. The strategy of having some work already completed to put into the proposal and using the funds for exploratory work does get around this limitation. If you believe that your most important responsibility is to do the best and most innovative science, then these minor deceits may seem justifiable. You may place them in the category of justifiable lies.

On the other hand these minor deceits can multiply. You may have to tell additional lies to cover up the first ones. From the perspective of the granting agency, you are committing fraud and may even be subject to legal action. The person or agency that is lied to usually views the deceit differently from the person who tells the lie.

The issue of peer review is discussed in detail by Daryl E. Chubin and Edward J. Hackett in *Peerless Science* (1990). The justification of lies is explored in *Lying* by Sissela Bok (1978).

CASE: BUYING A CALCULATOR

As a student you are on a very limited personal budget, but your research group has a very large budget for supplies. Next semester you will be taking a course in which a calculator with graphics capabilities will be required. These calculators are quite expensive, and it will be difficult for you to find that much extra money in your budget. While discussing your problem over lunch, a fellow student suggests that you charge the calculator to your advisor's research grant: "Dr. Morgan will never notice such a small amount of money. We spend that much on solvents every week in this group. Besides, you probably *will* also use it in your research from time to time." Should you take this advice?

Commentary: Buying a Calculator

This case raises the question of the proper use of research funds. Grants are usually awarded for quite specific purposes, and the expenditures are audited. Because academic research is quite free-ranging, all "reasonable" expenditures are allowed. The question, of course, is, What is reasonable? In this case the calculator can probably be justified as a research project expense because the student is likely to use it for some research-related work, but its primary purpose, at least initially, will be to perform course work. In addition, the student will treat the calculator as a personal possession, not as equipment belonging to the research group. When the student leaves the research group, he or she will probably take the calculator, not leave it behind for others to use.

Since it is a relatively small purchase it will probably not be noticed by the bookkeeper. Therefore it really is a personal moral question for the student and for the research advisor, who is ordinarily required to approve all purchases made on a grant. Technically, all purchases made with research grant funds are the property of the institution that accepts the grant. Practically, many small things, such as office supplies, effectively become personal property. Some, such as paper and pens, are used up. Others, staplers, scissors, and such, aren't worth worrying about. But where should the line be drawn? Taking a laptop computer purchased with grant funds would probably be regarded as theft. Is a $100 calculator in the same category?

It is best for those who manage research funds to set strict guidelines for their use. But if the guidelines are not clear, the student should certainly ask the research advisor for permission before making any questionable purchase.

CASE: RESEARCH FUNDS

You are a new graduate student in analytical chemistry working with Dr. Aston, a famous mass spectrometrist. Dr. Aston is paying you as a graduate research assistant using funds from an NIH grant that supports her work on the mass spectral identification of biological molecules such as DNA and proteins. When you begin work she asks you to study a series of organometallic cluster compounds which are important in industrial catalysis. None of these molecules have any direct biological significance.

She tells you that if the project works, she should be able to obtain contract funding from major oil companies, which is important now that government funds are tight. When you ask her whether it is acceptable for you to be paid from the NIH grant and work in a different area, she says, "Sure, everyone does it. Besides, if they ask, I can make up a good biochemical reason for your work. The project manager will never know the difference." Do you think this view is acceptable?

Commentary: Research Funds

This case raises an important issue. Can grant funds be used to support projects outside the scope of the proposal? Both federal and private granting agencies award grants based on fairly specific proposals. While the agencies understand that research is often unpredictable, they do expect that the actual work will conform fairly closely to the proposal.

On the other hand, if a researcher has a good idea she would like to explore prior to writing a major grant proposal or wants to pursue a fast-breaking development, where does the money for supplies and personnel come from? In some universities, seed money is available, but what if this source does not exist? How can interesting but controversial ideas be pursued? In his book *Polywater* (1981), Felix Franks speculates that much of the work on that phenomenon was supported by other grants. He uses the words *bootlegging* and *moonlighting*.

The grant process is slow. The time lag between submitting a proposal and receiving the funds is about twelve months. Since research moves much faster in some areas, the researcher's need for flexibility and spontaneity may conflict with her accountability to the granting agency. This conflict presents an ethical problem. Is the researcher's responsibility to the scientific enterprise more important than the responsibility to the granting agency?

Similar problems occur regularly. For example, funding for the salary for a graduate research assistant runs out before the student has finished the research and the writing of the dissertation, but there is money from another unrelated grant. Can the research advisor use the funds from the second grant to support the student for the last few months of his or her graduate career? In this example, the commitment to the education of the student conflicts with the strict provisions of the grant.

In deciding these questions, it is important to understand the overall mission of the granting agency and the specific terms of the grant or contract. Some granting agencies are very flexible; others are not. If the proposed use of the funds is within the spirit of the agency's mission and guidelines, then it is probably acceptable. When in doubt, a conversation with the program director for the granting agency should clarify the situation.

CASE: INDUSTRY-FUNDED ACADEMIC RESEARCH

You are doing graduate research with Dr. Giles, a bioanalytical chemist, studying the effects of compounds found in cigarette smoke and tar on biological samples. You feel that you are learning a great deal from your work, thoroughly enjoy what you are doing, and have developed a great relationship with Dr. Giles. After four months in the lab, you discover from another professor that Dr. Giles's research is primarily funded by a major tobacco corporation, though he does have a small grant from a private foundation. This information is interesting to you, but it is not very disconcerting because you have complete confidence in Dr. Giles. After another week of research, Dr. Giles informs you that the time has come for the two of you to write a paper concerning the results of the research you have done over the past months. He also emphasizes that the paper will be reviewed by the tobacco corporation before being submitted to the journal. In addition, you learn that Dr. Giles has a major consulting agreement with the corporation. These revelations concern you. You begin to wonder whether the connections with the tobacco company and the possible restrictions on publishing your work will affect your future career. What should you do?

Commentary: Industry-Funded Academic Research

You agreed to work with Dr. Giles to further your education, not to do research for a tobacco company. However, such industrial relationships are not generally detrimental to students' educations. In fact, industries often provide funding for academic research because the data obtained can be very useful to them. But, you still think that Dr. Giles is tangled in a conflict of interest here. He has a responsibility to the tobacco company because they are paying for the research and paying him for his knowledge, but he also has a responsibility to the academic and scientific community to provide unbiased, reliable data and information from the research findings. This is a delicate situation.

The essential question in this case is if the situation is undermining your education. Perhaps the best solution is to get a third opinion from someone trustworthy and not directly involved in the situation. This information may help you decide what to do. You can also wait a while after the paper is written to see if you think that the research is truly being affected negatively by Dr. Giles's involvement with the corporation.

In addition, you could present your feelings to the departmental administration, but this action comes with consequences. If they agree with you and take action against Dr. Giles, then the relationship with him would probably be destroyed. Plus, your research would come to a screeching halt. Even if they don't agree, in this situation, probably the best solution is to get a third opinion from someone not directly involved whom you trust. Ultimately, your action should be consistent with your personal values.

Some of the issues related to industry-funded academic research are discussed by Shulman (1999), Coppola (2001), Kovac (2001), McSherry (2001), and Djerassi (1993).

CASE: WHOSE NOTEBOOKS?

You have almost completed your Ph.D. in chemistry at a large research university. During your graduate career you did research under Dr. Ballok, a leading expert in bioanalytical chemistry, on the development of microsensors to monitor the concentrations of various substances in biological fluids. Dr. Ballok suggested that you work in this area and provided some initial guidance, but you have done most of the research independently. Although Dr. Ballok's research grants have supported the work, your results and ideas were an important part of his most recent successful renewal proposal. Not only has your work been considered an important advance in sensor design, it has considerable commercial potential.

As you are completing your dissertation, a midwestern university, known for its research in analytical chemistry, offers you a tenure-track position as an assistant professor, in part thanks to the glowing recommendation of Dr. Ballok. They have offered you a competitive salary and startup package, and you are anxiously anticipating the opportunity to continue your research at this university. As you are packing up your office, you go down to the lab to get your lab notebooks, which contain pages and pages of valuable information that you will need to begin your independent research program. On your way out of the lab, Dr. Ballok stops you and questions you about what you are doing. You explain to him that you need the information in the lab notebooks to continue your work on sensors in your new position. Dr. Ballok adamantly tells you that you may not take the information with you; the data you found have been funded by grants he wrote and therefore belong to his lab. In addition, Dr. Ballok and some of his graduate researchers plan to continue this research with the goal of starting a company that produces and markets biochemical sensors.

You argue that the data are yours; not only have you spent the past several years doing the experiments but most of the ideas were yours. Dr. Ballok refuses to compromise. He says that the lab notebooks belong to his lab, period.

Dr. Ballok leaves, and you become more furious with each passing moment. You are convinced that the lab notebooks belong to you and Dr. Ballok has absolutely no right to use the data that you have assiduously collected. Even more irritating is the suggestion that Dr. Ballock and his current students will be working to commercialize your discoveries. You know that if you actually take the notebooks, Dr. Ballok will be furious, your professional and personal relationship with him will be destroyed, and he might even press criminal charges. But there is a copy machine just down the hall where you could copy the data. Besides, Dr. Ballok did not mention anything about taking *copies* of the data. What should you do?

Commentary: Whose Notebooks?

This case raises the issue of ownership of data. You believe it is your right to take the data because you conducted all the experiments that produced it. Moreover, although Dr. Ballok did provide overall direction of the project, most of the ideas were yours. Dr. Ballok believes that because he wrote the grants that funded the research, the information belongs to his lab. He thinks that you have a responsibility to cooperate with him on this research done under his supervision. However, you think that because you did the actual data collection independently, the data belong to you.

In this case, no matter what, you must preserve your relationship with Dr. Ballok. He has been your mentor for the past five years, and it is because of him that you have this new, exciting job. You have a responsibility to respect your mentor and the university. It is a logical assumption that photocopying the data would, in the eyes of Dr. Ballok, be the same as taking the lab notebooks. Therefore you should not copy the lab notebooks. If you do photocopy the data and continue the research, there is a strong possibility that Dr. Ballok will find out and your reputation as a professional scientist could become that of a plagiarist or a thief. Your future as a respected scientist could be destroyed along with the possibility of future recommendations from Dr Ballok. You will be able to continue your research at the midwestern university for the general good and progression of science, but you will have to come up with new ideas. Actually, this is a good opportunity. Your new work will not be viewed as derivative from your graduate research. You will be able to develop a truly independent career.

However, Dr. Ballok did not behave perfectly. He should have informed you, preferably in writing at the beginning of your research in his lab, of his policies concerning the ownership of data. At the very least, he should have discussed this topic with you when you began applying for faculty positions. The two of you could have come to some agreement about which part of the research, if any, you would be permitted to continue and which part would remain with Dr. Ballok's group. He was the principal investigator on the projects, so he is entitled to keep the data and pursue the ideas developed in his group. His department and university will certainly expect him to maintain his distinguished research career. The students in his research group also deserve the opportunity to expand on the work done in that group. That is, after all, how research in chemistry works.

The potential commercial opportunities present another issue. On the one hand, it is true that Dr. Ballok is entitled to pursue the commercialization of his research, subject to the policies of the university and of the granting agency. The moral, and perhaps legal, question is whether, and if so to what extent, you should be included in this commercial opportunity. For example, if patent applications are filed, should you be listed as a coinventor? Although your expertise might be valuable, including you would mean that you should share in the financial rewards. These are difficult questions that Dr. Ballok should discuss openly with you, and he should justify his decision, whatever it is. Finally, if you disagree, can you appeal his decision?

CASE: YOUR DISCOVERY

You are working as an undergraduate research assistant in a physical chemistry laboratory, studying a class of compounds that have possible industrial applications as high-temperature materials. You are a very ambitious researcher and spend long nights in the lab. One night, after four months of work, you make a very important breakthrough: you synthesize a compound that is stable at very high temperatures, in excess of 1000 degrees centigrade. You immediately realize that if this compound is as useful as you hope, the financial rewards will be great.

The next day, you report your discovery to your supervisor, Dr. Walker. Of course he is very excited, but he informs you that you have no independent rights to the discovery. Because the research is sponsored by an industrial firm that has interests in the applications of product, the firm retains the primary right to license and use the compound. Moreover, because the work was done in the university lab, the school has the sole ownership of the patent of the compound. Dr. Walker says, "This sort of agreement is very common in research today."

You are furious; you have spent hours and hours over the past four months working with these compounds. Dr. Walker has had very little to do with the research. Besides, you had the idea to synthesize the particular compound that was particularly stable at very high temperatures. You decide that you are not going to let the industrial firm or Dr. Walker intimidate you. You tell yourself that it's your discovery. Tomorrow you will look into applying for a patent for the compound. Is this the right thing to do?

Commentary: Your Discovery

This case, based on a similar incident concerning undergraduate researcher, Petr Taborsky, who did research at the University of South Florida in the late 1980s, deals with the important issue of intellectual property rights. As described in Seth Shulman's book, *Owning the Future* (1999), Taborsky claimed that when he discussed filing for a patent after making a breakthrough, his supervising professor threatened him with jail. He was awarded two patents and then imprisoned for stealing university property because he filed the patents in his own name, not the university's. He got an early release from jail in the spring of 1997 (after serving eight weeks on a chain gang!) on good behavior. The university has sued him, claiming ownership of the patents, and Taborsky is on probation until 2008.

Approximately 7% of all scientific research is sponsored by private industry. In almost all cases, the corporation or firm legally reserves all licensing privileges and the university has all rights to patents. Even though Taborsky's case is extreme and even though he made the discovery, legally he stole intellectual property from the university.

Sponsor research leads to many scientific discoveries annually and undoubtedly contributes to the pool of knowledge for society, but researchers working under such a funding system must realize that they usually have no rights to the work. Such a system may seem to skew the fundamental ideals of independence in science and dampen the spark of innovation and competition, but because of the agreement made with the sponsor and the fact that the discovery was made in a university lab in an industry-sponsored university program, Taborsky legally committed theft of the university's property.

If you are working in such a program or have questions about any patent rights you might have in a university research lab, you should see your lab supervisor, department head, or someone knowledgeable in your institution's policies. Government agencies and other groups encourage, even more so after the Taborsky incident, universities to inform students whether they are hired as inventors or assistants and for institutions to post their policies about intellectual property rights in college catalogs and other easily accessible places. Before taking any kind of actions to patent a product you make in the lab, consult your university officials to see if such an action is legally permissible. Otherwise, you could land in jail and have your career ruined.

CASE: GRADUATING WITH HONORS

Your graduation is nearing, and your parents promised you a brand new car if you graduate with honors. Unfortunately, your GPA is right on the borderline; everything depends on a single grade, your current chemistry course. If you earn an A, you're in; with a B, it will be very close. You are sure of an A in all your other courses.

You are anxious about the chemistry course because the professor is planning to base the grade in the class on a curve and use a take-home final. The curve means that you must do well compared with all the other students; the students with the highest scores will receive As, but those at the bottom will receive Fs. Because you are willing to work as many hours as it takes, you think your chances of doing well in the course and graduating with honors are high.

Your professor has specified that although there is no time limit, the take-home final should be done individually. Your college has an honor code that governs behavior in such situations. You have been working on the final by yourself, but some of the questions are frustrating. A few even seem impossible. Jenny, your longtime friend who is also in the class, invites you to join her group to do the take-home final. When you inquire further, you find out that the whole class has divided into groups to complete the final. The instructor suspects nothing. You did not give Jenny a specific answer, but you promise to think about it. What should you do?

Commentary: Graduating with Honors

This case raises several issues. The professor has trusted the students, under the college honor code and as aspiring professional chemists. On the other hand, you have been promised a nice reward for an excellent record, a new car, along with the distinction of graduating with honors. There are several possible courses of action.

If you accept Jenny's suggestion, you will probably receive an A in the course and graduate with honors, not to mention earn the much-desired car from your parents. But if you take this route, you will be violating the college honor code, the professional code of chemistry, and, almost certainly, your personal moral standards. However, it appears as if the entire class has chosen to violate the trust of the instructor, which might make the action seem acceptable. Of course, the instructor might find out about the group work and assign you (and the rest of the class) a failing grade so you might not be able to graduate after all.

Another option is to refuse the offer and do the assignment by yourself, but say nothing to the instructor about the groups. Yet you know those who don't work in groups will be at a disadvantage. The whole class will probably do much better than you on this exam. Because the course is graded on a curve, your grade will probably suffer. The result may be that you will have a clear conscience but a poor grade. If the professor does discover the widespread collaboration, you will be innocent and will not suffer the consequences.

The third solution is not to accept the offer and to inform the professor. This action will almost certainly make you unpopular in the class. No one likes a whistleblower. You might be obligated under the honor code of your college to report academic misconduct, and as an aspiring scientist, you certainly have the responsibility to point out breaches of professional ethics to the appropriate authority, in this case the instructor. There are some personal benefits. Your own exam grade will look much better if the others are assigned an F. Your professor will probably be impressed with your moral courage. On the other hand, there is likely to be at least psychological retribution from the others in the class, which will make your final weeks in college much less enjoyable. And it is still possible that the professor will not judge that your performance on the final exam, and in the rest of the course, deserves an A, so you might also lose both the car and the honors distinction.

Although the third course of action seems the best in the abstract, in the real world these decisions are much more complicated.

CASE: A LONG LAB

Your organic lab has been working almost the entire period on a very difficult and time-consuming synthesis of a certain aromatic-based enzyme of great importance to the beef industry. Your professor, Dr. Riley, has told your class that this experiment demonstrates a common and important synthesis strategy and that a good working knowledge of this strategy will help you excel in later classes and in graduate school. You have been working very hard for the whole lab period, but you just cannot figure out how to complete one step that requires a 45% yield to provide enough intermediate for the next step to be successful. You are growing frantic, as are most of your classmates, who are having similar difficulties. Dr. Riley has told the class that this lab period will be the only one available to work on this experiment. Dr. Riley has agreed to keep the lab open for an extra few hours, and any extra time required to finish this week's enzyme synthesis will require you to stay after the regularly scheduled period. You have a big date planned for tonight, and if you can get out of lab on time, you will have just enough time to make it home and get ready for the evening.

A fellow professor comes into the lab and begins to talk to Dr. Riley. After a few minutes they both leave the room engaged in an animated conversation, which means that Dr. Riley might be gone for quite some time. Several students immediately begin to copy results from Cal, the class whiz, who has almost completed his synthesis. Cal asks if you want to copy, but you remember that at the beginning of the semester Dr. Riley told the class that any kind of cheating or forging of data would result in an automatic "F" for the course. Dr. Riley might return at any time, and Cal needs to turn in his lab notebook at the end of the period. Do you sacrifice your date to finish your work or copy from Cal at the risk of being caught?

Commentary: A Long Lab

The primary ethical question in this case is the dishonest acquisition of data for a class laboratory assignment. There are seemingly two courses of action. The first is to cheat while Dr. Riley is out of the room. Your personal frustration with the synthesis makes the temptation to copy appealing. Other factors that might influence you are the fact that a large number of other students in the lab are already copying Cal's lab notes along with Cal's willingness to provide the data. Peer pressure can be a powerful factor in decision-making processes. If everyone is doing it, it must be acceptable. Besides, Dr. Riley is likely to be gone for a while, so you are not likely to get caught. And if you are late for your date, or even worse, don't show up, your relationship could be ruined.

Despite these factors enticing you to copy from Cal, there are also numerous contrary considerations to weigh. First, there is the chance that Dr. Riley might come back in and catch people cheating. You must decide whether you want to take the chance of possibly ruining your future at this school by receiving the automatic *F*; in addition, you would earn the hard-to-remove stigma of unethical behavior. Another factor that lowers the appeal of cheating is your concern for your future knowledge and ability to perform synthesis reactions of this type, which are common and important. If you cheat now, will you know how to perform similar syntheses in future labs? Finally, and most importantly, cheating is an act that essentially every moral code condemns. Cheating in science is absolutely unacceptable.

Another option is to stay in the lab until you can figure out this problem and complete the synthesis. Although this approach may conflict with your date plans, you will avoid the risk of getting caught cheating and receiving an automatic *F*. And you will have the bonus of knowing how to perform the reaction, which you may use in graduate school and the job market.

Clearly the better choice in this matter is to accept the short-term inconvenience of having to stay late in the lab. It will save you from possible punishment, and having this knowledge will make you a better chemist. Most importantly, it will make you a better person and a responsible professional.

CASE: A CHALLENGING LAB REPORT

You are spending another late night. A lab report over your enzyme kinetics lab that you and your lab partners have just completed (for the second time!) is due the next day. Because none of the data obtained during the first lab period worked out for the calculations (the calculated values of the Michaelis-Menton constants were too low compared with the literature), your group had to redo the experiment and is spending a lot of time together to figure out the problem. You discover that the data you collected in the second experimental session give you numbers that are too high relative to the values that have been cited in the literature for comparison.

After comparing your group's data with those of another group and with the literature values, someone suggests early in the morning that you manipulate the data a little and use some of the numbers that the other groups measured to obtain calculated values closer to those cited in the literature. This suggestion is tempting to everyone, especially because it would simplify the writing of the lab report. Time is running short, and it would help everyone's grade. After all, the first run of the enzyme kinetics did not go well, and this disappointed the professor and put everyone behind schedule. Another problem is that there is no more enzyme to perform the experiments even if you did have time. Should you accept this suggestion?

Commentary: A Challenging Lab Report

Not having an experiment run successfully is frustrating, especially if your experiment has been repeated and still does not seem to work out. In the case of a lab experiment or a research project where you have strong expectations about the outcome, the temptation to massage the data to make sure that the results are "correct" is certainly strong. Data points can be dropped after using Q tests and other types of statistical analyses, but this does not guarantee that the data will turn out like you hope. The situation in this case study raises the issues of data manipulation and the dishonest acquisition of data.

Because this situation takes place in an undergraduate chemistry lab where the experiment is being done for a class, not for research purposes, manipulating the data may seem like a minor transgression. After all, who will be adversely affected? The only people involved are your group, the professor, the other lab groups in this class, and perhaps future classes. If the course grades are assigned based on an absolute scale, rather than a curve, then this small act of dishonesty will not adversely affect anyone's grade; it will only improve yours. Because it is likely that the professor will not notice the data manipulations, no one will ever know the difference.

By manipulating the data to obtain results closer to the accepted literature values, however, your group is being dishonest to the professor, others in the class, and the scientific process. Science requires what Richard Feynman (1985) called a kind of "utter honesty." Changing data, even in a situation where it doesn't really matter, is a bad habit to develop. There are also some more immediate consequences. If the professor were to find out that the group changed the data, he would certainly regard it as a form of cheating and impose a severe grade penalty.

There are several possible reasons for the unsuccessful experiment. These include having a bad sample that, in this case, may not have contained enough of the necessary enzyme; faulty equipment; and poor technique. One of the important learning objectives in any laboratory course is learning to identify the sources of error. It is far better for the group to be honest and try to determine the sources of error than to try to cover up the error by trimming the data. It is also important to recognize that all measurements involve error: random error, systematic error, or mistakes. Even if the final result does not agree precisely with a literature value, the experiment has not necessarily been done improperly. And even if there are significant errors, you might be able to identify potential problems with the experimental procedure, which the instructor can correct for future students. Moreover, presenting a careful discussion of the errors in your lab report may well improve your grade.

It is easy, and sometimes tempting, to manipulate data so that the expected results are obtained. Yielding to this temptation is a serious violation of any conception of scientific ethics.

The role of prior expectations in the interpretation of data is discussed in the context of the Millikan oil drop experiment by Holton (1978). For a general discussion see the Sigma Xi booklet, *Honor in Science* (1986).

CASE: PRESSURE ON A FRIEND

During your first year in graduate school at a major research university in the Northeast, you become friends with a shy but capable fellow student named Sam, who is from a small town in Idaho. Sam was an undergraduate at a small college in the West and is the first member of his family to graduate from college, let alone go on to graduate school. Sam has chosen to work with the rising star of the department, an ambitious faculty member just promoted to associate professor. Sam is a good research worker, and his advisor has asked him to work on his most speculative project, the one which, if successful, will probably make his reputation.

Over the summer after your first year in graduate school, Sam is, at first, very successful in his research. This success wins him lots of praise and additional responsibility. Things seem to deteriorate in July, however, as Sam is unable to complete a crucial step in the work. His advisor becomes impatient, and Sam puts in long hours trying to accomplish this step. By early August Sam is exhausted and becoming depressed.

In mid-August you take a week off to go home. You leave, quite worried about your friend, who is working too hard and beginning to drink too much. When you return, however, he reports that he has been successful and things are going well again. When you ask for details, he is strangely unwilling to share them. He just says, "I got lucky and things finally worked." His manner is very defensive. You begin to wonder whether he really did complete the step. What do you do?

Commentary: Pressure on a Friend

In this situation Sam is under enormous stress and you begin to suspect that he has cheated in some way to overcome his obstacle. Unfortunately, you have no hard evidence. If you raise your suspicions, you can damage Sam's credibility. If you do nothing and he has committed some sort of scientific misconduct, it is possible that he will get away with it, at least for a while. If he is caught later, the consequences will probably be worse. If he is not caught, then a piece of fraudulent research will be in the literature, perhaps leading others astray.

If you try to investigate further, Sam may think you are meddling in his affairs and become angry. This might destroy your relationship and further diminish his self-esteem. If you snoop around in his notebooks and other data, *you* might be accused of misconduct.

It seems that every course of action raises some sort of moral question. To make a good decision, it is important that you first try to clarify the facts as much as possible. Although you will need to be discreet, it is essential that you learn as much as possible about what Sam is doing. It may be that the facts will exonerate him. If, however, it appears that his actions are questionable, then you need to find the best course of action. You should try to find someone trustworthy with whom you can discuss the situation, preferably a senior faculty member who has some knowledge of professional ethics. Many institutions have a protocol for dealing with possible incidents of scientific misconduct. If yours does, you should follow it carefully. Such policies usually have procedural safeguards to protect the accused.

You also need to be aware that whistleblowers often become victims themselves, so you need to decide whether you are willing to take the personal risk that might be involved.

CASE: A TROUBLESOME ARTICLE

While reading an article published by Dr. Ross, your former postdoctoral advisor, you notice some disturbing things. The work being reported in this paper was going on while you were in his group. In fact, you helped one of the graduate students collect some of the spectroscopic data. The spectra published in the article lack one of the puzzling features of the spectra you helped obtain. You remember that there was a distinctive—and curious—line that now seems to be missing. There was much discussion of this line at the time. No one had a good explanation for it, which is why the data were not published then.

The absence of the line troubles you. One easy explanation is that it was due to an impurity, which, in the interim, they managed to remove. But you remember that herculean efforts had been made to purify the system and eliminate any interference and that Dr. Ross had concluded that the feature was real. The other possible explanation is more disturbing: the line was merely erased. During your time in Dr. Ross's lab, he was always scrupulously honest in his work, but in the intervening years his research program has fallen on hard times. In the difficult funding climate he has lost a major long-term grant, and he is under some pressure to rebuild his funding base. If you raise a question about this paper, his reputation will probably be damaged. On the other hand, you are suspicious. What should you do?

Commentary: A Troublesome Article

Here we raise the dilemma of the whistleblower. You have some reason to suspect scientific misconduct, but you can't be sure. If you ask Dr. Ross directly, he will probably be defensive or insulted. If you raise the issue in a more public way, you run the risk of damaging his reputation. Even a false allegation of misconduct will hurt. On the other hand, you have a responsibility to the scientific community to act on your suspicions in some way. If Dr. Ross has published a fraudulent paper, then something should be done about it. In thinking about this case, here are some questions to consider.

1. What are the facts of the case? What do you remember from your time in Dr. Ross's research group? Have scientific knowledge and technique progressed enough since that time to explain the difference? For example, have separation techniques been developed to remove the possible impurity? Are there experiments that you can do in your own lab to answer these questions?

2. What is your responsibility to the scientific community to report suspected cases of misconduct? If there is a problem with the paper, is it something that others working actively in this area will catch quickly? If the results are fraudulent, will the profession or the public suffer serious consequences?

3. How can one proceed in such cases while preserving the dignity of the suspect? What steps can be taken to make sure he is presumed innocent until proven guilty?

4. Is there someone at Dr. Ross's university that you can contact, in confidence, to obtain additional information about the situation? Can someone inform you of the institutional protocols for handling possible scientific misconduct?

5. In many whistleblowing situations, the accuser is attacked by the accused and his supporters. Should this be a consideration in your decision? Who will be affected if you are attacked?

CASE: A DIFFICULT ADVISOR

You are working at a famous research institute under a world-renowned scientist. As you begin your work, a major paper is being readied for publication. To prepare for your own project, you begin trying to reproduce some of the simpler findings of this paper. You work very hard to master the experimental techniques and repeat the experiments several times to make sure that you have not made any mistakes, but you are unable to reproduce the earlier work. Certain that you have discovered a problem, you go to talk to your advisor. Unfortunately, she is arrogant and difficult to work with. When you question the results of the research, she becomes angry at your insinuations. She then assigns you to a different and more menial job.

When searching for some background data one day, you happen upon the original notebooks of the experiment you had trouble reproducing. The results recorded in the notebooks do not match the published results. You notice the results of the control experiments seem to have been altered. The notebooks are sloppy, and it is difficult to follow the experimental procedures. As you read and understand the notebooks, the whole experiment seems to fall apart. If you are right, the major paper, which has just appeared in a leading journal, may be incorrect, even fraudulent. How do you proceed?

Commentary: A Difficult Advisor

The major issue is in this case is whistleblowing. You have uncovered evidence that scientific fraud may have been committed and you must decide how to proceed. The first complication is that you have a poor relationship with your famous advisor. She has rebuffed your earlier attempt to talk with her about your difficulties in reproducing the experimental results. Is she hiding something, is she just so arrogant that she cannot accept any criticism, or is there some other reason? Because you are less experienced, is it possible that you are in error? If you raise the issue of misconduct, it is likely that you will not be believed. Your advisor has a distinguished reputation. No one will want to believe that she has *committed* misconduct. If you do raise the issue, you may be in for a long and difficult fight. Your own integrity and motivations might well be challenged.

Questions to consider in planning a course of action include

1. What are the important moral principles that should govern your decision? One, of course, is your responsibility to maintain the integrity of science, but another is your need to protect your own self-interest, including your long-term job prospects. In this context, your personal relationships (spouse, children, parents, and others) might play a large role.

2. Because any accusation of misconduct might damage your advisor's reputation, how sure must you be before you raise the issue publicly? What are the consequences if you are wrong?

3. Strategically, what is the best way to proceed? Should you go to your advisor, another faculty member, the department chair, the dean, or an institutional scientific integrity committee? What are the moral and practical consequences of each of these decisions?

4. Will other people be implicated if you raise this issue? What consequences will they face?

In most institutions there is a well-defined procedure to follow in reporting possible cases of scientific misconduct. While the protocol may be clear, most whistleblowers have a difficult time ahead of them. The natural response of the professional community is to close ranks and defend itself. Often the whistleblower suffers personally. On the other hand, if you say nothing, an instance of scientific misconduct may go undetected, perhaps with serious consequences for both the scientific community and the public.

There is an extensive literature on whistleblowers. Some useful references include Callahan (1988, chap. 9), Davis (1998, chap. 6), Glazer and Glazer (1989), and Alford (2001). A detailed historical account of the case of alleged scientific misconduct involving Nobel laureate David Baltimore has been given by Kevles (1998).

CASE: TOO MANY SAMPLES

You have been Dr. Nutton's research assistant for several months and thus far have focused on collecting data for a grant application that is due at the end of the month. The assays you have been performing have taken much longer than anticipated, and the number of samples still to be analyzed is almost unmanageable. Attempts to discuss this fact with Dr. Nutton have been unsuccessful because of his insistence that everything be done as soon as possible. With only three weeks until the deadline, you begin to panic. It is Tuesday, and Dr. Nutton expects all the data by the end of the week so that he can get the application written up. But several hundred samples remain to be analyzed. You fear losing your job if you are unable to finish the work.

While reviewing the results of the completed experiments, in hope of noticing something that might help the work go faster, you notice a trend among the measurements taken from previous samples. It would be easy, you realize, to make up intelligible results for the remaining samples simply by using the measurements already taken. Dr. Nutton does not suspect any problem with your completing the work by Friday and therefore should not be suspicious of your results. With time running out, you decide to randomly select samples to analyze from those remaining and fill in the rest. You finish Thursday afternoon, and on Friday morning you give the results to Dr. Nutton, who is very pleased that everything is finished half a day early.

On Monday morning, Dr. Nutton comes to you very excited about the results. He informs you that the samples in the later groups were the ones of most concern, but because the results show the same trends, there should be no problem securing the grant, not to mention several publications from the upcoming work. He had not mentioned the differences in the samples because he didn't want you to be biased in your work in any way. Now he knows that the results are valid. Heading back to his office to work on the grant application, Dr. Nutton thanks and praises you for your work, mentioning that he had been a bit concerned about your performance in the beginning but that he has nothing to worry about now. You are immediately concerned about your fraudulent actions. What should you do?

Commentary: Too Many Samples

The issue in this case is not whether you should decide to fabricate data, but what you should do once you realize the implications of your fraudulent actions. What seemed to be an easy way out of a difficult situation has become a potentially serious problem.

There are at least two ways to proceed. You could immediately tell Dr. Nutton what you have done. This course of action might have serious consequences for you. Not only could you lose your job you might also lose your reputation and not be able to find a similar job again. This, of course, will depend on how Dr. Nutton responds to the news. It is possible that he might respond generously and understand the pressures that caused you to engage in dishonesty. On the other hand, his disappointment could lead to a swift and harsh reaction.

Another option is to go back to the lab and check some of the samples for which you fabricated results. It is certainly possible that your guesses or interpolations were actually correct, or at least nearly so. Although this does not excuse your earlier behavior, it might save the grant proposal and even your job. If you do find that your made-up data are far from the mark, you will have to go to Dr. Nutton and tell him the truth.

Clearly, you have made a mistake, but it is important to realize that pressure can tempt a technician or any other research worker to engage in a similar deception. It is also important for research directors to understand that putting such pressures on the junior members of their research teams can lead to misconduct, and it is essential for those doing the bench work to have the courage to tell the research director that the work simply cannot be done before the deadline. A humane and open atmosphere in the laboratory can prevent such incidents.

CASE: STAR POSTDOC

David Emory, a postdoctoral fellow, came into your research group with exceptional credentials and glowing recommendations. As a graduate student, he was coauthor of fifteen publications. His productivity has been no less spectacular during the year that he has worked in your lab; he has been a major contributor to four publications already. He is clearly brilliant, ambitious, and hard-working, and you have been recommending him for assistant professorships at major universities around the country.

At present Emory is working on an important problem in phase separation of polymer systems. This experiment requires that data be taken at regular intervals over long periods of time. Some of the systems require a week to come to equilibrium. A new postdoctoral fellow in the group, Jennifer Clark, comes to you and tells you that she has been suspicious of Emory's work for several months now because of the large number of experiments that he has performed. She accuses Emory of fabricating data. She claims to have seen him labeling data taken over only a period of a few hours with the incorrect times of "24 hours," "36 hours," "48 hours," and "60 hours."

When you confront Emory with the accusation, he denies it. He says that Jennifer Clark and the others in the group are jealous of his success and are not skilled enough to make a judgment about the experiments he was performing. He reminds you of his past accomplishments both as a graduate student and in your own group and of his growing national reputation. He dismisses the accusations as nonsense.

Emory's Ph.D. advisor is both an old and trustworthy friend and an eminent scientist. The work that you have published with Emory has been well received by the scientific community and is the basis for a major grant proposal that you have just submitted. What do you do about Jennifer Clark's accusations?

Commentary: Star Postdoc

In this case we approach the whistleblower issue from a different angle: your response to allegations of misconduct in your research group. Clearly, you have a lot invested in David Emory. You have published a number of papers with him and have strongly recommended him for jobs. His work is the basis of a major proposal. Consequently, if some of his work turns out to be fraudulent, your own reputation will suffer along with his. Since Emory has been a productive and seemingly trustworthy member of your research group, your natural tendency would be to dismiss Jennifer Clark's accusations. She is new to your group, and you don't know how much you can trust her. Perhaps her charges do stem from jealousy.

On the other hand, your reputation will suffer even more if the fraud persists and someone else exposes it. Although you don't know Jennifer very well, she did have the courage to talk to you about the alleged misconduct, and Emory's answers when you confronted him were a bit evasive. These facts put you in a difficult situation. In such cases the crucial issue is fairness to all concerned while discovering the truth. Here are some questions to consider:

1. Who will be affected if some of Emory's work is fraudulent? In what ways will they be affected?
2. What are the consequences to Jennifer Clark if her accusations are false?

3.   Can you, as research advisor, serve as an impartial judge of the facts in this case? If not, who else would be an appropriate judge?

4.   Should Emory be allowed to continue his work while the allegations are being investigated? Should he be more closely supervised?

5.   If the allegations prove to be true, what should you do about the publications, the grant application, and the letters of recommendation?

6.   Should you involve Emory's doctoral advisor in the investigation?

7.   What evidence will convince you that Emory is innocent?

CASE: THE MISSING LAB NOTEBOOK

You began working in Dr. Norcom's research lab at the same time as his other sophomore assistant, Jerry. Because of some cutbacks in the chemistry department's research funds, Dr. Norcom now can pay only one assistant. Neither Jerry nor you can afford to work in the lab without pay. When you propose that each of you work half-time, Dr. Norcom will not agree. He believes that having two half-time assistants would be an inefficient use of his research funds. Since both of you are qualified and both need the job, he is uncomfortable deciding between you. To make everything fair, Dr. Norcom proposes that both of you work independently for the next two weeks and the student who makes the most progress will keep the job.

The lab is open all day so students can work at their convenience. Today is Tuesday of the second week, and Jerry mentions that he cannot find his lab book. Dr. Norcom does not know that Jerry is missing his lab book, but he would probably be angered by Jerry's lack of caution. Dr. Norcom has repeatedly warned the two of you to take care of your lab books and never leave them unattended. Shortly after Jerry leaves the lab for lunch, you discover that his lab book has fallen behind a desk. Do you retrieve the lab book? Do you keep it? Do you give the lab book personally to Dr. Norcom or to Jerry, or do you just lay it on the desk for someone to find? What is the best course of action?

Commentary: The Missing Lab Notebook

The question is, How serious a transgression has Jerry committed? If he is a less capable research assistant for having lost his notebook, it is important that Dr. Norcom know this. On the other hand, you might judge either that losing the lab notebook is really a minor problem or that values such as compassion and charity toward Jerry are more important than any responsibility you might feel to report his transgression.

The best course of action is to give the lab book to Jerry and not mention his carelessness to Dr. Norcom. Jerry has made a mistake by misplacing his lab book, but hardly a mistake that he should be severely punished for. Dr. Norcom is quite serious about taking care of lab notes; telling him about the incident would certainly hurt Jerry's chances to keep the job.

There is certainly a temptation for you either to give the lab book to Dr. Norcom and tell him where you found it or to give it to Jerry but still let Dr. Norcom know about Jerry's carelessness. These actions would increase your chances of keeping the paid research position. Another option is to leave the lab book behind the desk for someone else to find. You may feel you are avoiding any involvement, but in fact you are making a moral decision, a decision not to help Jerry or Dr. Norcom. It is also a decision to help yourself, because leaving the book behind the desk will certainly impede Jerry's research progress and probably make sure he does not get the permanent job. Yet another option would be to wait a few days to "find" the book and return it to Jerry. This would probably give you a time advantage, although Jerry might be able to continue his work without the lab book for a while, recording his results in a new notebook. Once it was returned, he could transfer his lab notes into the permanent record and proceed as if nothing happened.

Both of these options hurt Jerry's chances to retain the job, thus making yours better. From a purely selfish perspective they seem attractive. After all, only one person can have the job. Why not you? There is, of course, the issue of fairness. Dr. Norcom is interested in hiring the better researcher. Keeping track of lab notes is a part of research, but only a minor part. The more important qualities are good laboratory technique and scientific knowledge and insight: hands and head. Both Dr. Norcom and the scientific community will benefit most if the truly better research gets the position. From a utilitarian perspective, the best option is quietly to return the lab book to Jerry and let Dr. Norcom make his decision at the end of the week. This more generous course of action also makes sense from a variety of other moral perspectives. Certainly, if you had misplaced your notebook, you would like Jerry to return it to you without telling Dr. Norcom about your carelessness.

Several other options are posed in the case. You could retrieve the lab book and lay it out somewhere for anyone to find, offering Jerry and Dr. Norcom the same opportunity to find the lab book. If Dr. Norcom finds the lab book, carelessly lying out while Jerry is not in the lab, then Dr. Norcom will probably be angry and will want to hire you rather than Jerry. Although this option seems to absolve you of any responsibility if Jerry is reprimanded or punished for losing his lab book, it isn't quite that simple. You have a variety of choices as to where to leave the lab book and when. Consciously or unconsciously, you can certainly set up the situation so that Dr. Norcom is more likely than Jerry to find the lab book, or the reverse.

Finally, and most extreme, you could discard the lab book. While this would almost guarantee you the job, the notebook certainly contains information that is important to the overall research project, so you will be doing a disservice to science, no matter who eventually keeps the research assistant position.

CASE: DANGEROUS WASTE

Cindy's company, Sinistex Chem, regularly dumps a chemical into a local river. In preliminary tests this chemical shows a strong possibility of being a carcinogen. However, the management of Sinistex refuses to stop the dumping because altering the process would be costly and no regulations on the release of this particular chemical have been set by government agencies. Cindy has already thought of a way to remove the chemical and dispose of it safely, but whenever she brings it up with her supervisor, he tells her that if they remove the chemical now, before everyone has to remove it, it will put the company at a competitive disadvantage. What should Cindy do?

Commentary: Dangerous Waste

Cindy believes that the risk to the public from the release of the toxic substance is sufficiently great that the company should act. She has gone further and developed a process to remove the offending chemical from the waste stream. The company, however, disagrees. The company's position could be based solely on economic concerns (it is too expensive to remove the chemical from the waste) or on a different assessment of the risk to the public. Based on the information given in the case, it appears that the company is primarily concerned with cost.

First, Cindy needs to determine how serious the public risk really is. Evaluating risk is complicated (Rodricks 1992). The relative toxicity is important, of course, but questions such as dose and residence time also come into play. Because reducing the amount of the carcinogen costs money, the company will legitimately compare the cost to the benefit of introducing the process that Cindy has developed.

That being said, there is probably a difference in values. Sinistex seems more concerned with its competitive position than with public safety and is willing to wait until the substance in question is regulated. Cindy sees a threat and wants to act immediately.

The practical question is how Cindy should proceed. First, she needs to decide how important this battle is. In the end she might lose her job or decide that she needs to resign in protest, so the stakes could be high. Assuming that she feels strongly about the danger, her first step is to work within the company for a realization that it is in the company's interests to implement her solution. This means quantifying the risk and evaluating the benefit of the solution as well as its cost. Cindy needs to assemble all the relevant data on the health effects of the substance in question as well as an estimate of the costs that the company might incur if they continue to release the compound in their waste stream. People whose health is affected might well bring lawsuits. She should also compile information on the regulatory situation. The compound is currently unregulated, but it might be on a list of substances under consideration by the appropriate federal agency. If regulation is coming, the company might be persuaded to act sooner rather than later. Finally, the company might realize a public relations benefit in eliminating a carcinogen. Presenting this information to the Sinistex management might be persuasive.

If Sinistex is still unmoved, then Cindy must decide what to do next. Should she go to the appropriate federal agency and try to persuade it to regulate the substance? As an employee of Sinistex she might not be free to do this, so such an action could result in punishment by the company. She could also work through a local or national environmental organization, but with the same risks. Finally, she could "go public" herself, revealing the health risks in a press conference. This action would certainly cause her to lose her job at Sinistex.

Here the professional responsibility of a scientist or engineer comes into conflict with the values of the employer. Sadly, such ethical conflicts can result in personal disaster for the professional who is forced to become a whistleblower (Alford 2001).

CASE: SUPERVISOR IN TRAINING

Gene has just been hired in a new job in which he will oversee five chemists who carry out a complicated and potentially dangerous chemical reaction. The training for this job includes four weeks working with an experienced manager of the same process at the main plant before moving to another location where this process is also being run. His trainer, Jill, was amazed at how quickly he was learning how to deal with the problems that commonly arose. Even though she commented to her superiors that Gene could probably run everything just fine after two weeks, they had to continue for four according to company policy. At the start of the fourth week, however, Jill became sick and needed to go home, but no other people qualified to manage the reaction were available to fill in for her. She really needed to go home and rest, but she toughed it out for a day and barely made it to work the next.

"I can't keep working. I need to go home," begged Jill. "Gene, you can handle this for a day, and no one will know. Just say that I've stepped out for a few minutes if anyone asks. You know how to do this job. Don't worry about it." Should Gene allow Jill to go home? If not, what are his options?

Commentary: Supervisor in Training

Gene is being asked to take on a professional responsibility for which he is not formally qualified and to tell a white lie to cover for his trainer who is ill and needs to go home. This situation raises questions of both professional and personal ethics. Even though the four-week apprenticeship with an experienced manager is only an internal company requirement, not a professional certification, it was probably instituted for good reasons. Jill may be confident of Gene's abilities, but he is not certified. For Gene to supervise by himself is analogous to practicing law or medicine without a license. Perhaps a better analogy is driving or flying without a license. If everything goes well, no one will know the difference. If there is a serious problem, both Gene and Jill will be in trouble.

The first task is to identify all of Gene's options. To do this we need to know more about the structure and policies of the company. From the case statement, it appears that there is no other qualified manager available. Perhaps company protocol will allow Jill's supervisor or some one else, to act as manager in an emergency. If not, does the process have to be shut down? If so, what does this do to Gene's training program?

If Jill goes home leaving Gene in charge, he is put in a difficult position. If he reports to the management that Jill has left, he opens her up to reprimand. In addition, this action may destroy his good relationship with Jill, making the rest of his training period tense. She may retaliate by suggesting to management that her opinion of him has changed and that he cannot handle his new responsibilities. If he does what she suggests and lies to cover up her absence, the deceit might well be discovered, putting both of them in a difficult position.

It is irresponsible for Jill to leave Gene in a supervisory role for which he is not formally qualified. The best option is for the two of them to go to a supervisor, explain the situation, and work out a solution. If, however, Jill does leave, then it seems best for Gene to go to the supervisor himself regardless of the consequences to Jill. Both honesty and safety suggest that this is his best course of action.

CASE: RECOMMENDING A FRIEND

Your roommate, Julius, is a fellow chemistry major and friend from high school who came to your university after his junior year at a small college back home. He transferred because he wanted some big-school exposure and because he thought that recommendations from professors at a nationally recognized school would improve his chances of attending a prestigious graduate program. One night while Julius and a classmate study for a biology test, you hear Julius ask the classmate if he may copy from him during the test because he is not prepared. When you casually confront Julius about cheating later that week, he reveals that he does it all the time and says, "What is the point of memorizing things that I will never use in the future and I would probably forget no matter how much effort I put into it?" He does say that he rarely cheats in classes that interest him, especially chemistry.

Later that month, a professor whom you admire and trust and for whom you have done research asks you to comment on your friend Julius. It seems that Julius is doing well in the professor's chemistry class and has asked him to write a letter of recommendation to a graduate program in a well-known university. Because the professor knows of your friendship with Julius and because he respects your opinion, he has come to you to get a better idea of Julius's character and responsibility. What do you tell him?

Commentary: Recommending a Friend

Although you do not want to violate the trust of the professor who has asked you for an opinion, you also want to preserve your long-standing friendship with Julius. In addition, you do not want to hinder his application to graduate school. But you also realize that Julius is cheating not only himself but other students as well by achieving good grades through academic dishonesty.

If you tell the professor that Julius is of questionable character, you are fulfilling an obligation to be honest with someone who respects your opinion and trusts you. In addition, you are upholding the important principle of academic integrity. If your university has an honor code, you might be obligated under that code to report cheating. But reporting Julius's transgression means that he will probably get a poor recommendation from your professor, which may ruin his chances of being admitted to the graduate program. If he discovers that you informed the professor of his cheating, it could ruin your friendship.

If you decide to ignore the cheating and tell the professor that Julius is a great person, he will almost assuredly write Julius a glowing letter of recommendation, which means that Julius will probably be admitted to the graduate program. As far as you know his success in the professor's class has been earned honestly, and so the professor can truly say that he is bright and that his performance in chemistry has been excellent. But in this case, you would be ignoring important moral principles. The professor did ask for your opinion of Julius's overall character, not simply his intelligence and performance in chemistry. If Julius goes on to graduate school and performs poorly or, worse, is caught cheating, the professor's assessment of Julius's character will appear naïve and maybe even untruthful. By protecting your friend you may injure the professor you respect.

Another option is to talk to Julius about the situation and tell him how you feel about his cheating, giving him a chance to explain his behavior. You can decide on your next step based on what he says. He may be able to convince you that the cheating was an isolated incident and that his moral character is better than you fear, or you may conclude that he is a habitual cheater who can't really be trusted. You should probably inform him of what you plan to say to your professor, even at the risk of losing his friendship. Your honesty might just be what he needs to rethink his view of academic integrity. If you are really his friend, you should realize that you aren't doing him any favors by supporting his bad habits.

Stephen L. Carter (1996, chap. 5) has written eloquently about the ethical problems involved in the writing of recommendations.

CASE: ADVANCED LABORATORY PROJECT

Marshall, Paul, Jane, and Bev are students in a senior-level laboratory course in which students work in teams on open-ended projects of their own design. They are meeting to plan their report on a project in which they investigated the effect of solvent quality on the conformation of polymer chains using intrinsic viscosity as the experimental probe of the polymer dimensions.

Marshall, the team leader, opens the meeting by saying, "Last week we divided the write-up into four parts, and each of us agreed to bring a draft of our part to this meeting. My assignment was the theoretical background. I spent a lot of time trying to understand Flory's *Principles of Polymer Chemistry*, and the more modern stuff in DeGennes's *Scaling Concepts in Polymer Physics*, and I think I've put together a pretty good five-page summary of the main ideas. I have it all in Microsoft Word on this disk so we can put it into the final report. Jane and Bev were supposed to write up the experimental procedure and work up the data. How has that gone?"

"We worked on it together," Jane replied. "We have written up the experimental procedure. I think it's ready to include in the report. We put all the data into an Excel spreadsheet to work it up, but we have had some trouble getting the calculations right. We found all the constants in the *Polymer Handbook* and used them to calculate the chain dimensions, but something is wrong. The final numbers are weird. Bev and I were going to work on this some more tonight. Do you think that we could meet again tomorrow sometime? I hope by then we will have this figured out. If not, maybe you can take a look at what we've done."

"I could meet anytime tomorrow morning," replied Marshall. "I have something else to do in the afternoon." He then turned to Paul, "Paul, your job was to look up the theory of viscosity measurements. How has that gone?"

Looking a bit embarrassed, Paul replied, "I haven't had a lot of time this week, but I do have a couple of pages of handwritten notes. I found a pretty good source, *Polymer Chemistry: The Basic Concepts*, by Paul Hiemenz, but I had a lot of trouble understanding the mathematics of fluid flow. It sounds like what all of you have done will really impress Dr. Jones, so my part won't matter too much."

"But, Paul, this is a group project. We're all supposed to contribute more or less equally. We have until Monday. You still have time to finish your part." responded Bev.

"I don't really have time," answered Paul. "I have a major paper for my religious studies course also due Monday. Anyway, all of you are much better chemistry students than I am. You know I'm finishing the chemistry major just to satisfy my father. I'm applying to divinity schools for next fall. This religion paper is very important. If I get an *A* in the course, I might get into Harvard. Can't you three just use what I have done so that I can concentrate on this other assignment. Just do me a small favor."

"Okay, Paul, go work on your paper, and we'll finish the report. Leave me your notes and the Hiemenz book, and I'll finish up your part," said Marshall in a rather resigned tone.

After Paul had left, Bev asked, "Have any of you ever worked in a group with Paul before? Does he always do this?"

Marshall replied, "I've worked with him in a lot of groups. He does this about half the time; the excuses are always different, however. Unfortunately, if we want a good grade on this project, we need to do most of his part. Let's get started."

Consider the following questions:

1. Should Marshall, Jane, and Bev do the work assigned to Paul and turn in the report with all four names attached? What other options do they have?

2. What could the group have done at the beginning of the project to prevent this situation?

3. Should the three group members consult Dr. Jones? What responsibilities does she have to make sure that groups function properly?

4. If the situation cannot be resolved equitably, should the final grades be affected, and if so, how?

Commentary: Advanced Laboratory Project

Cooperative learning and team projects are increasingly a part of both undergraduate and graduate education in chemistry. This case raises a practical and ethical question that often arises in this context: how to deal with a team member who is does not fulfill his or her obligations. Paul has not completed his part of the report, and the three other members of the group have to decide how to handle the situation. The questions point to several options.

One option is for Marshall, Bev, and Jane to pick up the slack and write Paul's portion of the report. Because the report has to be written and all three of them want to receive credit—and a good grade—this seems to be the only realistic possibility; they will be able to turn in a complete report. If they choose this path, they face a second question. Should they inform Dr. Jones that Paul has not contributed? If they do, Paul's grade will suffer, perhaps ruining his chances to go on to divinity school even if he does well in the religious studies course. If they don't, Paul will receive credit for work he has not done. Although being honest about Paul's lack of participation seems to be the right thing to do, it isn't easy to inform on a classmate who might be a close personal friend. Assuming that Paul's story is true, the other students probably do not want to jeopardize his future. If he is planning to go to divinity school, what difference does it make that he doesn't understand polymer chemistry?

Another option is to turn in a partial report accompanied by an explanation of why Paul's part is missing. Although this is an honest option, there are certainly some risks. Dr. Jones might not look favorably on the partial report on the grounds that the assignment was to complete the project.

A third option is to ask Dr. Jones to intervene. He might be willing to talk to Paul and encourage, or force, him to finish his part of the report. This will probably affect Paul's grade in the course. In addition, Paul will probably be resentful and not do a very good job on the report, lowering the overall quality of the final product. Although Dr. Jones does have, in some sense, the responsibility to ensure that the group functions well, little can be done with a student who refuses to do the work. Ultimately, the proper functioning of teams depends on all members behaving responsibly.

CASE: AN OPINIONATED PROFESSOR

You are an African-American undergraduate student taking an organic chemistry class. You are having some trouble with the class and are in a dilemma. To stay in school, you must keep your scholarship, which requires you to maintain a GPA of 3.25 in all science courses. Your grade in organic chemistry is currently in doubt, but you are working to the best of your ability to meet the requirements of the scholarship.

Over the course of the semester, the professor has made several racial slurs insinuating that African-Americans are less intelligent than others. Because of these remarks you have been afraid to talk to the professor about your problems in the class. When the last test was returned, you noticed that there were some points taken off that you felt were not justified. These points are important because of your borderline grade in the class. When you try to talk to the professor, he dismisses you by saying, "Whatever the grade is on the paper is what you earned." How do you handle this situation?

Commentary: An Opinionated Professor

This case brings to the surface the problem of discrimination in the classroom. As in any case of sex, religion, or national origin discrimination, there are at least three possible routes of action. The first is to confront the professor. However, you have already done this and have been unsuccessful. To be successful, you could further research your "errors" and find evidence (preferably from your textbook, lecture notes, or lab notes) that you are indeed correct and the grading was unfair. In addition, you could compare your returned exam with those of fellow classmates to see if points were taken off their exams for similar mistakes. If they were not, you then have very persuasive evidence for the professor to change the grade. If points were subtracted from other students' exams for essentially the same answers, then you do not have a case.

Demonstrating racial discrimination can be difficult. Although you feel that the professor's actions have been discriminatory, it is possible that he is just being arbitrary. If you are in a large class, he may not have known who you were when he graded the exam, and his refusal to admit an error could just be a defensive response. While he may be racially insensitive, or worse, your grade may not be the result of discrimination. Before proceeding, make sure that you have a good case. You should also be prepared for the emotional toll that bringing such a case might have on you.

But confronting the professor on *this* point is dangerous. He may retaliate in various ways, including assigning you a poor grade. He might harass you verbally by increasing the number or the nastiness of the racial slurs. A sympathetic and knowledgeable faculty member or the campus office concerned with diversity may be able to assist you in assessing the situation and preparing your arguments and may also have good suggestions on how to proceed if your appeal to the professor is unsuccessful.

A second course of action would be simply to give up and suffer the loss of points that you think you earned. But you need every point that you can earn to keep your GPA at 3.25 and your scholarship money. What is more important, however, is that you seem to have been treated unfairly based only on race. If there has been discrimination, it is both illegal and unethical. Unless someone speaks up, the discrimination will continue.

If you do have a good case, you should pursue it, the third option. The appropriate course of action is to go to the department chair, the dean of students, or the campus office concerned with diversity affairs. Before you go, make sure you have collected the appropriate evidence to prove your case. In preparing your case, you might seek advice from someone with experience in dealing with issues of race discrimination, perhaps even a lawyer. It is possible that the person you talk to will give the benefit of the doubt to the professor, so you need to be well prepared. But if your case is strong, one of these offices will have the power to deal with the professor's behavior. Finally, if you are unsatisfied with the response of the university, it is always possible to take legal action.

CASE: LAB TECHNICIAN

You are a research technician working in Dr. Harris's lab. Last year, you graduated from the university with a B.S. in chemistry with the intention of going to medical school, but since you were not accepted, you have decided to take a year to explore the possibility of pursuing a graduate degree in chemistry. Because of several comments that have been made since your arrival and because you are the only woman in Dr. Harris's lab, you begin to believe that you were hired simply because of university regulations requiring an increase in the number of women employees, not because Dr. Harris thinks that you can adequately perform the necessary work. Your responsibilities have not been, in your opinion, appropriate for your ability, and you do not believe that the work is going anywhere. In addition, you have overheard Dr. Harris disparaging the abilities of women scientists to some of the other technicians in the lab. Though concerned about how Dr. Harris will respond, you decide to discuss the matter with him.

When you confront him, Dr. Harris tells you that he has recently received another major grant and there are several new projects that need to be started. He briefly describes the various projects, then gives you several background articles concerning the one he has assigned to you. While reading them, however, you notice that several of the reagents you will be using have been shown to affect the female reproductive system and it is not recommended that women work with them.

The following day, you confront Dr. Harris with this information, and he says that if you do not think that you can perform the work, he'll give it to one of the male technicians in the lab and you can just continue working as you have been. When you point out that there are other new projects, he tells you that those are too important and that he has already assigned them to the male technicians. When you try to explain to him that you can perform the research for any of the projects that he described as well as the other technicians, he dismisses you by saying that you have all the responsibilities that you can handle. What should you do?

Commentary: Lab Technician

This case brings up the issue of discrimination in the workplace. Although the specific laws and institutional policies that cover discrimination in the classroom and in the workplace may differ, the general principles are the same. You should refer to the commentary for An Opinionated Professor as you think about this problem.

Though you have confronted Dr. Harris with the fact that your duties have been menial, you have not yet confronted him with your belief that his decisions are gender-related. Because of the outcome of the discussion, however, you are now almost certain that Dr. Harris is making judgments about you based on your gender and not your ability. In your opinion, if he were not discriminating against you, he would have immediately understood the dangers you might face if you engaged in the project he originally assigned to you and switched you to another project. His response suggests that he was fully aware of the dangers involved and suggested the project *expecting* that you would turn down the offer so that he could keep you in a more menial position.

Allegations of sex discrimination are controversial because they put one person's feelings and perceptions against another's. They are difficult to prove and can often leave the accuser, even if a case is won in court, with a damaged reputation. The accused can often walk away with others perceiving him or her as the victim and not be affected much at all.

How should you proceed? You must first decide if you have a case for sex discrimination. What leads you to believe that you do or do not? How do you decide if you have a case? Are there people you can talk to about it, either in the department or outside of it, say the campus diversity affairs office, the office of human resources, a lawyer, or some other person?

After you have an idea where you stand, if you feel that you do have a case, what should you do next? Do you tell Dr. Harris what you suspect? Is it possible that he will be willing to make some changes if he is aware of your feelings? Is that going to be enough for you? Should it be? Do you complain to the department or other university officials and let them investigate? Do you file a lawsuit against him? When?

If you are concerned that you do not have a case, you must also decide what you should do. Do you make Dr. Harris aware of your concerns at this point? Do you just keep quiet and continue your job as before? Should you discuss the problem with another faculty member, maybe a female faculty member in your own or another department?

Whatever course of action you take, you must consider the emotional and professional effects on you. To what extent, if any, should you also consider how it will affect Dr. Harris or the department? Clearly, there is no one best way to proceed.

Although this case is written as a gender-discrimination issue, the same concerns apply to other forms of discrimination that are found in the workplace as well as in the classroom. And although the scene in this specific case is a university laboratory, the same problems occur in private industry or government laboratories.

CASE: A HARMLESS JOKE

You and your friend Angela are seniors working for Dr. Burell, a prestigious chemistry professor. You grade papers, work as teaching assistants in the teaching labs, and conduct research for Dr. Burell, who is also your biochemistry professor. One day while working in the lab with the two of you, Dr. Burell tells a joke that both you and Angela consider highly offensive to women, yet you both politely laugh and continue to work. For the rest of the week, Dr. Burell continues to tell offensive jokes that he says he finds on the Internet. Angela has told you she feels very uncomfortable working with Dr. Burell when he tells her the jokes. You agree with Angela, and you tell her that Dr. Burell's jokes are a form of sexual harassment, which should not be taken lightly. Angela is understandably apprehensive about confronting Dr. Burell because he has a legendary short temper and a large ego. Questioning his actions might upset him. Angela fears that her job as well as her grade in the biochemistry class will be jeopardized if she confronts Dr. Burell. Angela needs this job experience and a good grade in Dr. Burell's class to get accepted into graduate school. To protect her future, she asks you to keep the incidents quiet. You know that Dr. Burell is committing a form of sexual harassment, which is against both school policy and the law, but you do not want to jeopardize your friend's future. What should you do?

Commentary: A Harmless Joke

In this situation, a person's ability to work and gain an education is hindered by sexual harassment. Sexual harassment of students includes unwelcome sexual advances, requests for sexual favors, and/or other verbal or physical conduct of a sexual nature where grades or educational progress are made contingent upon submission to such conduct, or when the conduct has the purpose or effect of interfering with the individual's academic performance, or of creating an intimidating, hostile, or offensive educational environment. Sexual harassment of employees is similarly prohibited. Sexual harassment is illegal, so Dr. Burell's actions are against the law as well as being immoral.

An important consideration is that Angela's future as a student and a chemist might be jeopardized if these incidents are reported because Dr. Burell might retaliate in some way. The dilemma in this situation is whether you should report the sexual harassment, which is ethically and legally wrong, or keep Angela's career safe by keeping quiet. After all, reporting this case might hurt Dr. Burell's career in chemistry.

An obvious solution to this problem would be to confront Dr. Burell. If he realizes that his actions are inappropriate, he might apologize, change his behavior, and resolve the whole situation. However, considering Dr. Burell's temper and ego, this approach might worsen the situation. He may also become mad at you for questioning his actions and authority, and your grade could suffer. Your case will be stronger if you mention that Angela is also offended by his jokes, but if you do, Dr. Burell might also become angry at her. Even if you don't say anything about Angela, Dr. Burell might assume that she agrees with you. This is precisely what Angela wants to avoid. She needs the work experience and a high grade to get into graduate school. Angela may also feel betrayed and be angry with you for going against her wishes.

A second option is to keep quiet about the whole situation, protecting both Dr. Burell and Angela's career. The problem remains that illegal actions would go unreported. Angela would continue working under uncomfortable conditions which could affect her performance. Other women in the department might also be sexually harassed by Dr. Burell if the department is not notified. It is also possible that Dr. Burell's behavior might be part of a general pattern within the department or the university that persists because the victims have been unwilling to speak up.

A third solution is to go directly to the dean of students, the department of human resources, or the department chair and have one of them handle the situation. Every college or university is required to have a published procedure for dealing with such complaints. It is probably posted and available in the student and employee handbooks. With the administration watching, it will be difficult for Dr. Burell to retaliate by assigning a lower grade. Dr. Burell will probably be reprimanded, which may keep the problem from growing and preserve the reputation of the department. The department could recognize internal problems and deal with them properly.

In unequal power relationships it can be dangerous to bring inappropriate behavior to the attention of the authorities. It is important to have good evidence. In cases like this it is common for the person in authority to side with the faculty member, dismissing the student's complaints as either false or an overreaction. Before you report Dr. Burell's behavior it would be wise to have a confidential conversation with the dean of students or the Office of Diversity Affairs or with someone else knowledgeable about sexual harassment to make sure that you have all the necessary evidence and that you have framed your complaint effectively.

CASE: JOB CANDIDATE

Your department is searching for a faculty member in the area of materials chemistry. All professors in the department are expected to compete successfully for external research funds. The candidate visiting today has very impressive academic credentials and some interesting ideas concerning his future research direction. When asked about possible sources of funding for his work, he makes it clear that he will not accept money from any defense agencies. He explains that he is a member of a traditional peace church and his personal values will not allow him to be involved with the military. This position clearly puts him at a disadvantage in obtaining funding for his work. Should it be a consideration in the discussion of his suitability for the appointment?

Commentary: Job Candidate

In this case, the tension is between the candidate's personal values and an aspect of his research program, the possible sources of funding. The same kind of tension can occur in other ways. You might discover that your research had potentially disastrous environmental consequences or that a compound you just made is an excellent chemical warfare agent. Although these issues occur more often in the biomedical sciences, they are not uncommon in chemistry.

The job candidate portrayed in this case has his own values sorted out. The question is how to be fair to him and to the department. Your department will invest both time and money in him in the hopes that he will become a successful professor. Setting him up for failure would not be fair to him or to your department. On the other hand, his credentials and ideas indicate that he has the potential to become a creative and productive scientist. If he can obtain adequate funding from nonmilitary sources, then his personal values are irrelevant. Some will argue that they are irrelevant in any case because everyone makes choices about where to apply for grants. Are his self-imposed restrictions any reason not to give him the chance to succeed? Does his willingness to disclose his restriction provide an argument for hiring him?

Gender, racial and ethnic diversity, and disability are related issues. One traditional argument to exclude women and minorities has been that they will have much more trouble succeeding because of prejudice (everyone else's, not ours) and therefore we ought not take the chance. The same argument is made about those with physical handicaps: "A blind person can't succeed in this field." A job candidate's qualifications cannot simply be laid out along a linear scale. A wide variety of factors must be considered.

CASE: PERSONAL VALUES

The brightest of the entering graduate student class has agreed to work with you for her Ph.D. You have a very interesting project, supported by the Army chemical warfare service, in mind for her to work on. Although the compounds are prospective chemical warfare agents, they also have interesting chemistry. The project will be difficult, and you are delighted to have such an excellent student to work on it. You describe the project to her, emphasizing the interesting chemistry and the generous funding. She tells you that she will not work on a project related to chemical warfare because it is against her personal values. How do you respond?

Commentary: Personal Values

This case raises the same issue as the Job Candidate case the tension between personal values and research. Here the student does not want to be involved in a project related to chemical warfare, despite its other attractions. As her advisor you do not share these values, but you do want to have her work in your group. The challenge for you is to affirm her values and to find her a project that she will be willing to work on.

Appendix

Codes of Ethics

I. AMERICAN CHEMICAL SOCIETY (ACS): THE CHEMIST'S CODE OF CONDUCT*

The American Chemical Society expects its members to adhere to the highest ethical standards. Indeed, the federal Charter of the Society (1937) explicitly lists among its objectives "the improvement of the qualifications and usefulness of chemists through high standards of professional ethics, education and attainments. . . ."

Chemists have professional obligations to the public, to colleagues, and to science. One expression of these obligations is embodied in "The Chemist's Creed," approved by the ACS Council in 1965. The principles of conduct enumerated below are intended to replace "The Chemist's Creed." They were prepared by the Council Committee on Professional Relations, approved by the Council (March 16, 1994) and adopted by the Board of Directors (June 3, 1994) for the guidance of society members in various professional dealings, especially those involving conflicts of interest.

Chemists Acknowledge Responsibilities To:

The Public

Chemists have a professional responsibility to serve the public interest and welfare and to further knowledge of science. Chemists should actively be concerned with the health and welfare of coworkers, consumers, and the community. Public comments on scientific matters should be made with care and precision, without unsubstantiated, exaggerated, or premature statements.

The Science of Chemistry

Chemists should seek to advance chemical science, understand the limitations of their knowledge, and respect the truth. Chemists should ensure that their scientific contributions, and those of the collaborators, are thorough, accurate, and unbiased in design, implementation, and presentation.

The Profession

Chemists should remain current with developments in their field, share ideas and information, keep accurate and complete laboratory records, maintain integrity in all conduct and publications, and give due credit to the contributions of others. Conflicts of interest and scientific misconduct, such as fabrication, falsification, and plagiarism, are incompatible with this code.

---

*Reprinted with permission of the American Chemical Society. © 2002 by the American Chemical Society.

The Employer

Chemists should promote and protect the legitimate interests of their employers, perform work honestly and competently, fulfill obligations, and safeguard proprietary information.

Employees

Chemists, as employers, should treat subordinates with respect for their professionalism and concern for their well-being, and provide them with a safe, congenial working environment, fair compensation, and proper acknowledgment of their scientific contributions.

Students

Chemists should regard the tutelage of students as a trust conferred by society for the promotion of the students' learning and professional development. Each student should be treated respectfully and without exploitation.

Associates

Chemists should treat associates with respect, regardless of the level of their formal education, encourage them, learn with them, share ideas honestly, and give credit for their contribution.

Clients

Chemists should serve clients faithfully and incorruptibly, respect confidentiality, advise honesty, and charge fairly.

The Environment

Chemists should understand and anticipate the environmental consequences of their work. Chemists have responsibility to avoid pollution and to protect the environment.

II. AMERICAN CHEMICAL SOCIETY: ETHICAL GUIDELINES OF CHEMICAL RESEARCH*

The guidelines embodied in this document were revised by the Editors of the Publications Division of the American Chemical Society in January 2000.

Preface

The American Chemical Society serves the chemistry profession and society at large in many ways, among them by publishing journals which present the results of scientific and engineering research. Every editor of a Society journal has the responsibility to establish and maintain guidelines for selecting and accepting papers submitted to that journal. In the main, these guidelines derive from the Society's definition of the scope of the journal and from the editor's perception of standards of quality for scientific work and its presentation. An essential feature of a profession is the acceptance by its members of a code that outlines desirable behavior and specifies obligations of members to each other and to the public. Such a code derives from a desire to maximize perceived benefits to society and to the profession as a whole and to limit actions that

*Reprinted with permission from "Ethical Guidelines to the Publication of Chemical Research," *Chem. Rev.* 2001, 101, 13A–15A. © 1985, 1989, 2001 by the American Chemical Society.

might serve the narrow self interests of individuals. The advancement of science requires the sharing of knowledge between individuals, even though doing so may sometimes entail forgoing some immediate personal advantage. With these thoughts in mind, the editors of journals published by the American Chemical Society now present a set of ethical guidelines for persons engaged in the publication of chemical research, specifically, for editors, authors, and manuscript reviewers. These guidelines are offered not in the sense that there is any immediate crisis in ethical behavior, but rather from a conviction that the observance of high ethical standards is so vital to the whole scientific enterprise that a definition of those standards should be brought to the attention of all concerned. We believe that most of the guidelines now offered are already understood and subscribed to by the majority of experienced research chemists. They may, however, be of substantial help to those who are relatively new to research. Even well-established scientists may appreciate an opportunity to review matters so significant to the practice of science.

Guidelines

A. Ethical Obligations of Editors of Scientific Journals

1. An editor should give unbiased consideration to all manuscripts offered for publication, judging each on its merits without regard to race, religion, nationality, sex, seniority, or institutional affiliation of the author(s). An editor may, however, take into account relationships of a manuscript immediately under consideration to others previously or concurrently offered by the same author(s).

2. An editor should consider manuscripts submitted for publication with all reasonable speed.

3. The sole responsibility for acceptance or rejection of a manuscript rests with the editor. Responsible and prudent exercise of this duty normally requires that the editor seek advice from reviewers, chosen for their expertise and good judgment, as to the quality and reliability of manuscripts submitted for publication. However, manuscripts may be rejected without review if considered inappropriate for the journal.

4. The editor and members of the editor's staff should not disclose any information about a manuscript under consideration to anyone other than those from whom professional advice is sought. (However, an editor who solicits or otherwise arranges beforehand, the submission of manuscripts may need to disclose to a prospective author the fact that a relevant manuscript by another author has been received or is in preparation.) After a decision has been made about a manuscript, the editor and members of the editor's staff may disclose or publish manuscript titles and authors' names of papers that have been accepted for publication, but no more than that unless the author's permission has been obtained.

5. An editor should respect the intellectual independence of authors.

6. Editorial responsibility and authority for any manuscript authored by an editor and submitted to the editor's journal should be delegated to some other qualified person, such as another editor of that journal or a member of its Editorial Advisory Board. Editorial consideration of the manuscript in any way or form by the author-editor would constitute a conflict of interest, and is therefore improper.

7. Unpublished information, arguments, or interpretations disclosed in a submitted manuscript should not be used in an editor's own research except with the consent of the author. However, if such information indicates that some of the editor's own research is unlikely to be profitable, the

editor could ethically discontinue the work. When a manuscript is so closely related to the current or past research of an editor as to create a conflict of interest, the editor should arrange for some other qualified person to take editorial responsibility for that manuscript. In some cases, it may be appropriate to tell an author about the editor's research and plans in that area.

8. If an editor is presented with convincing evidence that the main substance or conclusions of a report published in an editor's journal are erroneous, the editor should facilitate publication of an appropriate report pointing out the error and, if possible, correcting it. The report may be written by the person who discovered the error or by an original author.

9. An author may request that the editor not use certain reviewers in consideration of a manuscript. However, the editor may decide to use one or more of these reviewers if the editor feels their opinions are important in the fair consideration of a manuscript. This might be the case, for example, when a manuscript seriously disagrees with the previous work of a potential reviewer.

B. Ethical Obligation of Authors

1. An author's central obligation is to present an accurate account of the research performed as well as an objective discussion of its significance.

2. An author should recognize that journal space is a precious resource created at considerable cost. An author therefore has an obligation to use it wisely and economically.

3. A primary research report should contain sufficient detail and reference to public sources of information to permit the author's peers to repeat the work. When requested, the authors should make a reasonable effort to provide samples of unusual materials unavailable elsewhere, such as clones, microorganism strains, antibodies, etc., to other researchers, with appropriate material transfer agreements to restrict the field of use of the materials so as to protect the legitimate interests of the authors.

4. An author should cite those publications that have been influential in determining the nature of the reported work and that will guide the reader quickly to the earlier work that is essential for understanding the present investigation. Except in a review, citation of work that will not be referred to in the reported research should be minimized. An author is obligated to perform a literature search to find, and then cite, the original publications that describe closely related work. For critical materials used in the work, proper citation to sources should also be made when these were supplied by a nonauthor.

5. Any unusual hazards inherent in the chemicals, equipment, or procedures used in an investigation should be clearly identified in a manuscript reporting the work.

6. Fragmentation of research reports should be avoided. A scientist who has done extensive work on a system or group of related systems should organize publication so that each report gives a well rounded account of a particular aspect of the general study. Fragmentation consumes journal space excessively and unduly complicates literature searches. The convenience of readers is served if reports on related studies are published in the same journal, or in a small number of journals.

7. In submitting a manuscript for publication, an author should inform the editor of related manuscripts that the author has under editorial consideration or in the press. Copies of those manuscripts should be supplied to the editor, and the relationships of such manuscripts to the one submitted should be indicated.

8. It is improper for an author to submit manuscripts describing essentially the same research to more than one journal of primary publication, unless it is a resubmission of a manuscript rejected for or withdrawn from publication. It is generally permissible to submit a manuscript for a full paper expanding on a previously published brief preliminary account (a "communication" or "letter") of the same work. However, at the time of submission, the editor should be made aware of the earlier communication, and the preliminary communication should be cited in the manuscript.

9. An author should identify the source of all information quoted or offered, except that which is common knowledge. Information obtained privately, as in conversation, correspondence, or discussion with third parties, should not be used or reported in the author's work without explicit permission from the investigator with whom the information originated. Information obtained in the course of confidential services, such as refereeing manuscripts or grant applications, should be treated similarly.

10. An experimental or theoretical study may sometimes justify criticism, even severe criticism, of the work of another scientist. When appropriate, such criticism may be offered in published papers. However, in no case is personal criticism considered to be appropriate.

11. The co-authors of a paper should be all those persons who have made significant scientific contributions to the work reported and who share responsibility and accountability for the results. Other contributions should be indicated in a footnote or an "Acknowledgments" section. An administrative relationship to the investigation does not of itself qualify a person for co-authorship (but occasionally it may be appropriate to acknowledge major administrative assistance). Deceased persons who meet the criterion for inclusion as co-authors should be so included, with a footnote reporting date of death. No fictitious name should be listed as an author or co-author. The author who submits a manuscript for publication accepts the responsibility of having included as co-authors all persons appropriate and none inappropriate. The submitting author should have sent each living co-author a draft copy of the manuscript and have obtained the co-author's assent to co-authorship of it.

12. The authors should reveal to the editor any potential conflict of interest, e.g., a consulting or financial interest in a company, that might be affected by publication of the results contained in a manuscript. The authors should ensure that no contractual relations or proprietary considerations exist that would affect the publication of information in a submitted manuscript.

C. Ethical Obligations of Reviewers of Manuscripts

1. Inasmuch as the reviewing of manuscripts is an essential step in the publication process and therefore in the operation of the scientific method, every scientist has an obligation to do a fair share of reviewing.

2. A chosen reviewer who feels inadequately qualified to judge the research reported in a manuscript should return it promptly to the editor.

3. A reviewer (or referee) of a manuscript should judge objectively the quality of the manuscript, of its experimental and theoretical work, of its interpretations and its exposition, with due regard to the maintenance of high scientific and literary standards. A reviewer should respect the intellectual independence of the authors.

4. A reviewer should be sensitive to the appearance of a conflict of interest when the manuscript under review is closely related to the reviewer's work in progress or published. If in

doubt, the reviewer should return the manuscript promptly without review, advising the editor of the conflict of interest or bias. Alternatively, the reviewer may wish to furnish a signed review stating the reviewer's interest in the work, with the understanding that it may, at the editor's discretion, be transmitted to the author.

5. A reviewer should not evaluate a manuscript authored or co-authored by a person with whom the reviewer has a personal or professional connection if the relationship would bias judgment of the manuscript.

6. A reviewer should treat a manuscript sent for review as a confidential document. It should neither be shown to nor discussed with others except, in special cases, to persons from who specific advice may be sought; in that event, the identities of those consulted should be disclosed to the editor.

7. Reviewers should explain and support their judgments adequately so that editors and authors may understand the basis of their comments. Any statement that an observation, derivation, or argument had been previously reported should be accompanied by the relevant citation. Unsupported assertions by reviewers (or by authors in rebuttal) are of little value and should be avoided.

8. A reviewer should be alert to failure of authors to cite relevant work by other scientists, bearing in mind that complaints that the reviewer's own research was insufficiently cited may seem self-serving. A reviewer should call to the editor's attention any substantial similarity between the manuscript under consideration and any published paper or any manuscript submitted concurrently to another journal.

9. A reviewer should act promptly, submitting a report in a timely manner. Should a reviewer receive a manuscript at a time when circumstances preclude prompt attention to it, the unreviewed manuscript should be returned immediately to the editor. Alternatively, the reviewer might notify the editor of probable delays and propose a revised review date.

10. Reviewers should not use or disclose unpublished information, arguments, or interpretations contained in a manuscript under consideration, except with the consent of the author. If this information indicates that some of the reviewer's work is unlikely to be profitable, the reviewer, however, could ethically discontinue the work. In some cases, it may be appropriate for the reviewer to write the author, with copy to the editor, about the reviewer's research and plans in that area.

11. The review of a submitted manuscript may sometimes justify criticism, even sever criticism, from a reviewer. When appropriate, such criticism may be offered in published papers. However, in no case is personal criticism of the author considered to be appropriate.

D. Ethical Obligations of Scientists Publishing Outside Scientific Literature

1. A scientist publishing in the popular literature has the same basic obligation to be accurate in reporting observations and unbiased in interpreting them as when publishing in a scientific journal.

2. Inasmuch as laymen may not understand scientific terminology, the scientist may find it necessary to use common words of lesser precision to increase public comprehension. In view of the importance of scientists' communicating with the general public, some loss of accuracy in that sense can be condoned. The scientist should, however, strive to keep public writing, remarks, and interviews as accurate as possible consistent with effective communication.

3. A scientist should not proclaim a discovery to the public unless the experimental, statistical, or theoretical support for it is of strength sufficient to warrant publication in the scientific literature. An account of the experimental work and results that support a public pronouncement should be submitted as quickly as possible for publication in a scientific journal. Scientists should, however, be aware that disclosure of research results in the public press or in an electronic database or bulletin board might be considered by a journal editor as equivalent to a preliminary communication in the scientific literature.

III. AMERICAN INSTITUTE OF CHEMISTS: CODE OF ETHICS*[†]

The profession of chemistry is increasingly important to the progress and the welfare of the community. The chemist is frequently responsible for decisions affecting the lives and fortunes of others. To protect the public and maintain the honor of the profession, the American Institute of Chemists has established the following rules of conduct.

It Is the Duty of the Chemist

1.   To uphold the law, not to engage in illegal work nor cooperate with anyone so engaged;
2.   To avoid associating or being identified with any enterprise of questionable character;
3.   To be diligent in exposing and opposing such errors and frauds as the Chemist's special knowledge brings to light;
4.   To sustain the institute and burdens of the community as a responsible citizen;
5.   To work and act in a strict spirit of fairness to employers, clients, contractors, employees, and in a spirit of personal helpfulness and fraternity toward other members of the chemical profession;
6.   To use only honorable means of competition for professional employment; to advertise only in a dignified and factual manner; to refrain from unfairly injuring, directly or indirectly, the professional reputation, prospects, or business of a fellow Chemist, or attempting to supplant a fellow chemist already selected for employment; to perform services for a client only at rates that fairly reflect costs of equipment, supplies, and overhead expenses as well as fair personal compensation;
7.   To accept employment from more than one employer or client only when there is no conflict of interest; to accept commission or compensation in any form from more than one interested party only with the full knowledge and consent of all parties concerned;
8.   To perform all professional work in a manner that merits full confidence and trust; to be conservative in estimates, reports, and testimony, especially if these are related to the promotion of a business enterprise or the protection of the public interest, and to state explicitly any known bias embodied therein; to advise client or employer of the probability of success before undertaking a project;
9.   To review the professional work of other chemists, when requested, fairly and in confidence, whether they are: (a) subordinates or employees, (b) authors of proposals for grants

---

*Reprinted with permission of the American Institute of Chemists.
[†]Approved by the AIC Board of Directors, April 29, 1983.

or contracts, (c) authors of technical papers, patents, or other publications, or (d) involved in litigation;

10.   To advance the profession by exchanging general information and experience with fellow Chemists and by contributing to the work of technical societies and to the technical press when such contribution does not conflict with the interests of a client or employer; to announce inventions and scientific advances first in this way rather than through the public press; to ensure that credit for technical work is given to its actual authors;

11.   To work for any client or employer under a clear agreement, preferably in writing, as to the ownership of data, plans, improvements, inventions, designs, or other intellectual property developed or discovered while so employed, understanding that in the absence of written agreement: (a) results based on information from the client or employer, not obtainable elsewhere, are the property of the client or employer; (b) results based on knowledge or information belonging to the Chemists, or publicly available, are the property of the Chemist, the client or employer being entitled to their use only in the case or project for which the Chemist was retained; (c) all work and results outside of the field for which the Chemist was retained or employed, and not using time or facilities belonging to a client or employer, are the property of the Chemist; and (d) special data or information provided by a client or employer, or created by the Chemist and belonging to the client or employer, must be treated as confidential, used only in general as part of the Chemist's professional experience, and published only after release by the client or employer;

12.   To report any infractions of these principles of professional conduct to the authorities responsible for enforcement of applicable laws or regulations, or to the Ethics Committee of The American Institute of Chemists, as appropriate.

References

Alford, C. F. 2001. *Whistleblowers: Broken Lives and Organizational Power*. Ithaca: Cornell University Press.

Baird, D. 1997. Scientific instrument making, epistemology, and the conflict between gift and commodity economies. *Technè: Electronic Journal of the Society for Philosophy and Technology* 2(3–4): 25–46. Available at http://scholar.lib.vt.edu/ejournals/STP/stp.html

Barber, B. 1952. *Science and the Social Order*. New York: Free Press.

Barden, L. M., P. A. Frase, and J. Kovac. 1997. Teaching scientific ethics: A case studies approach. *American Biology Teacher 59*: 12–14.

Bayless, M. D. 1981. *Professional Ethics*. Belmont, CA: Wadsworth Publishing.

Beauchamp, T. L. 1991. *Philosophical Ethics: An Introduction to Moral Philosophy*. 2d ed. New York: McGraw-Hill.

Beauchamp, T. L., and J. F. Childress. 2001. *Principles of Biomedical Ethics*. 5th ed. Oxford: Oxford University Press.

Bebeau, M. J., K. D. Pimple, K. M. T. Muskavitch, S. L. Borden, and D. H. Smith. 1995. *Moral Reasoning in Scientific Research: Cases for Teaching and Assessment*. Bloomington, IN: Poynter Center.

Bell, R. 1992. *Impure Science*. New York: J. Wiley & Sons.

Blackburn, S. 2002. *Being Good: A Short Introduction to Ethics*. New York: Oxford University Press.

Bok, S. 1978. *Lying*. New York: Pantheon Books.

———. 1995. *Common Values*. Columbia, MO: University of Missouri Press.

Broad, W., and N. Wade. 1982. *Betrayers of the Truth*. New York: Simon & Schuster.

Brock, W. H. 1992. *The Norton History of Chemistry*. New York: W. W. Norton.

Bronowski, J. 1956. *Science and Human Values*. New York: Harper Torchbooks.

Brown, G. E., Jr. 1992. The objectivity crisis. *Am. J. Phys.* 60: 779–781.

Bush, V. [1945] 1990. *Science: The Endless Frontier*. Washington, DC: National Science Foundation.

Callahan, J. C., ed. 1988. *Ethical Issues in Professional Life*. New York: Oxford University Press.

Carter, S. L. 1996. *Integrity*. New York: Basic Books.

Christensen, C. R., D. A. Garvin, and A. Sweet, eds. 1991. *Education for Judgment: The Artistry of Discussion Leadership*. Boston: Harvard Business School Press.

Chubin, D. E., and E. J. Hackett. 1990. *Peerless Science*. Albany: State University of New York Press.

Close, F. 1991. *Too Hot to Handle: The Race for Cold Fusion*. Princeton: Princeton University Press.

Committee on Assessing Integrity in Research Environments. 2002. *Integrity in Scientific Research: Creating an Environment that Promotes Responsible Conduct*. Washington, DC: National Academy Press.

Committee on Professional Training. 1999. *Undergraduate Professional Education in Chemistry: Guidelines and Evaluation Procedures*. Washington, DC: American Chemical Society.

Committee on Science, Engineering and Public Policy. 1995. *On Being a Scientist: Responsible Conduct in Research.* 2d ed. Washington, DC: National Academy Press.

Coppola, B. P. 2000. Targeting entry points for ethics in chemistry teaching and learning. *J. Chem. Educ.* 77: 1506–1511.

———. 2001. The technology transfer dilemma: Preserving morally responsible education in a utilitarian entrepreneurial academic culture. *Hyle: Intl. J. for Phil. of Chem.* 7: 155–167.

Coppola, B. P., and D. H. Smith. 1996. A case for ethics. *J. Chem. Educ.* 73: 33–34.

Cournand, A., and M. Meyer. 1976. The scientist's code. *Minerva* 14: 79–96.

Croll, R. P. 1984. The noncontributing author: An issue of credit and responsibility. *Perspectives in Biology and Medicine* 27:401.

Davis, M. 1982. *Business and Professional Ethics Journal* 1: 17–28.

———. 1987. The moral authority of a professional code. In *Authority revisited, Nomos XXIX*, edited by J. R. Pennock and J. W. Chapman, 302–337. New York and London: New York University Press.

———. 1990. The ethics boom: What and why. *The Centennial Review* 34: 163–386.

———. 1993. Ethics across the curriculum: Teaching professional responsibility in technical courses. *Teaching Philosophy* 16: 205–235.

———. 1995. Ethics in engineering and technology: What to teach and why. *J. Tenn. Acad. Sci.* 70: 55–57.

———. 1998. *Thinking Like an Engineer.* New York: Oxford University Press.

———. 1999. *Ethics and the University.* London and New York: Routledge.

Djerassi, C. 1991. *Cantor's Dilemma.* New York: Penguin Books.

———. 1993. Managing competing interests: Chastity vs. promiscuity. In *Ethics, Values and the Promise of Science*, edited by Sigma Xi, 31–45. Research Triangle Park, NC: Sigma Xi.

Dyson, F. J. 1993. Science in trouble. *The American Scholar* 62: 513–522.

Feynman, R. P. 1985. *Surely You're Joking, Mr. Feynman!* New York: W. W. Norton.

Franks, F. 1981. *Polywater.* Cambridge: MIT Press.

Frase, P. A., L. M. Barden, and J. Kovac. 1997. *Scientific Ethics for High School Students.* Madison, WI: Institute for Chemical Education.

French, A. P. 1979. *Einstein: A Centenary Volume.* Cambridge: Harvard University Press.

Gass, W. 1980. The case of the obliging stranger. In *Fiction and the Figures of Life*, edited by W. Gass, 225–241. Boston: Godine.

Gert, B. 1988. *Marality: A New Justification of the Moral Rules.* New York: Oxford University Press.

Glazer, M. P., and P. M. Glazer. 1989. *The Whistleblowers: Exposing Corruption in Government and Industry.* New York: Basic Books.

Goodstein, D. 1991. Scientific fraud. *The American Scholar* 60: 505–515.

Gratzer, W. 2000. *The Undergrowth of Science: Deception, Self-Deception and Human Frailty.* Oxford: Oxford University Press.

Gross, A. G. 1996. *The Rhetoric of Science.* Cambridge: Harvard University Press.

Guston, D. H. 1999. *Between Politics and Science: Assuring the Integrity and Productivity of Research.* Cambridge: Cambridge University Press.

Guston, D. H., and K. Keniston. 1994. Introduction: The social contract for science. In *The Fragile Contract: University Science and the Federal Government*, edited by D. H. Guston and K. Keniston Cambridge: MIT Press.

Hardwig, J. 1985. Epistemic dependence. *J. Phil.* 82:335–349.

———. 1991. The role of trust in knowledge. *J. Phil.* 88:693–708.

———. 1994. Toward and ethics of expertise. In *Professional Ethics and Social Responsibility*, edited by D. E. Wueste, 83–101. Lanham, MD: Rowman & Littlefield.

Harris, C. E., Jr., M. S. Pritchard, and M. J. Rabins. 1996. *Engineering Ethics: Concepts and Cases*. Belmont, CA: Wadsworth Publishing.

Hermes, M. E. 1996. *Enough for One Life Time: Wallace Carothers Inventor of Nylon*. Washington, DC: American Chemical Society & Chemical Heritage Foundation.

Hinde, R. A. 2002. *Why Good Is Good: The Sources of Morality*. London: Routledge.

Hoffmann, R. 1988. Under the surface of the chemical article. *Angew. Chem. Int. Ed. Engl.* 27:1593.

———. 1997. Mind the shade. *Chem. Eng. News* (10 November):3.

Holton, G. 1978. Sub-electrons, presuppositions, and the Millikan-Ehrenhaft dispute. In *The Scientific Imagination*, edited by G. Holton. Cambridge: Cambridge University Press.

———. 1994. On doing one's damnedest: The evolution of trust in science. In *The Fragile Contract*, edited by D. H. Guston and K. Keniston, 59–81. Cambridge: MIT Press.

Hounshell, D. A., and J. K. Smith. 1988. *Science and Corporate Strategy*. Cambridge: Cambridge University Press.

Hyde, L. 1979. *The Gift: Imagination and the Erotic Life of Property*. New York: Vintage.

Johnson, D. G., and H. Nissenbaum, eds. 1995. *Computers, Ethics and Social Values*. Englewood Cliffs, NJ: Prentice-Hall.

Jonsen, A. R., and S. Toulmin. 1988. *The Abuse of Casuistry: A History of Moral Reasoning*. Berkeley and Los Angeles: University of California Press.

Kevles, D. J. 1998. *The Baltimore Case: A Trial of Politics, Science and Character*. New York: W. W. Norton.

Kitcher, P. 2001. *Science, Truth and Democracy*. Oxford: Oxford University Press.

Knight, D. 1992. *Ideas in Chemistry*. New Brunswick, NJ: Rutgers University Press.

Knight, D., and H. Kragh, eds. 1998. *The Making of the Chemist*. Cambridge: Cambridge University Press.

Kovac, J. 1995. *The Ethical Chemist*. Knoxville: University of Tennessee.

———. 1995. Ethics in physical science. *J. Tenn. Acad. Sci.* 70:53–54.

———. 1996. Scientific ethics in chemical education. *J. Chem. Educ.* 73:926.

———. 1997. Conflict of interest in chemistry. *Perspectives on the Professions*, Illinois Institute of Technology 17:6–7.

———. 1998. The ethical chemist. *Council on Undergraduate Research Quarterly* (March), 109–113.

———. 1999. Professional ethics in the college and university science curriculum. *Science and Education* 8:309–319.

———. 2000. Professionalism and ethics in chemistry. *Found. Chem.* 2:207–219.

———. 2000. Science, law, and the ethics of expertise. *Tenn. Law Review* 67:397–408.

———. 2001. Gifts and commodities in chemistry. *HYLE: Intl. J. for Phil. of Chem.* 7:141–153.

Kovac, J., and B. P. Coppola. 2000. Universities as moral communities. *Soundings: An Interdisciplinary Journal* 83 (3–4): 765–777.

Kuhn, T. S. 1962. *The Structure of Scientific Revolutions*. Chicago: University of Chicago Press.

LaFollette, M. C. 1992. *Stealing into Print*. Berkeley and Los Angeles: University of California Press.

Langmuir, L. 1989. Pathological science. *Phys. Today* 42(7):36–48.

Locke, D. 1992. *Science as Writing*. New Haven: Yale University Press.

MacIntyre, A. 1966. *A Short History of Ethics*. New York: Macmillan Publishing.

Macrina, F. L. 1995. *Scientific Integrity: An Introductory Text with Cases*. Washington, DC: ASM Press.

McGrayne, S. B. 2001. *Prometheans in the Lab*. New York: McGraw-Hill.

McSherry, C. 2001. *Who Owns Academic Work: Battling for Control of Intellectual Property*. Cambridge: Harvard University Press.

Medawar, P. B. 1964. Is the scientific paper fraudulent? *Saturday Review* (1 August 1964):42–43.

Merton, R. K. 1973. The normative structure of science. In *The sociology of science*, edited by R. K. Merton, 267–278. Chicago: University of Chicago Press.

Moore, A. D. 1982. Henry A. Rowland. *Scientific American* 246(2):150–161.

Nelkin, D. 1984. *Science as Intellectual Property*. New York: Macmillan Publishing.

———. 1987. *Selling Science*. New York: W. H. Freeman.

Nye, M. J. 1980. N-rays: An episode in the history and psychology of science. *Hist. Stud. Phys. Sci.* 11:125–156.

Panel on Scientific Responsibility and the Conduct of Research. 1992. *Responsible Science*. Vol. 1. Washington DC: National Academy Press.

Panel on Scientific Responsibility and the Conduct of Research. 1993. *Responsible Science*. Vol. 2. Washington, DC: National Academy Press.

Penslar, R. L., ed. 1995. *Research Ethics: Cases and Materials*. Bloomington, IN: Indiana University Press.

Polanyi, M. 1964. *Personal Knowledge*. New York and Evanston: Harper Torchbooks.

Popper, K. 1965. *The Logic of Scientific Discovery*. New York: Harper & Row.

Proctor, R. N. 1991. *Value-Free Science*. Cambridge: Harvard University Press.

Rachels, J. 1999. *The Elements of Moral Philosophy*. 3d ed. Boston: McGraw-Hill.

Reagan, C. E. 1971. *Ethics for Scientific Researchers*. 2d ed. Springfield, IL: Charles C Thomas.

Reese, K. M., ed. 1976. *A Century of Chemistry*. Washington, DC: American Chemical Society.

Resnik, D. B. 1999. *The Ethics of Science: An Introduction*. London: Routledge.

Reuben, J. A. 1996. *The Making of the Modern University*. Chicago: University of Chicago Press.

Rodricks, J. V. 1992. *Calculated Risks*. Cambridge: Cambridge University Press.

Rosenfield, S., and N. Bhushan. 2000. Chemical synthesis: complexity, similarity, natural kinds, and the evolution of a logic. In *Of Minds and Molecules: New Philosophic Perspectives in Chemistry*, edited by N. Bhushan and S. Rosenfeld, 187–207. Oxford: Oxford University Press.

Shapin, S. 1989. Who was Robert Hooke? In *Robert Hooke: New Studies*, edited by M. Hunter and S. Schaffer, 253–285. Woodbridge, Suffolk: Boydell Press.

———. 1994. *A Social History of Truth*. Chicago: University of Chicago Press.

Shulman, S. 1999. *Owning the Future*. Boston: Houghton Mifflin.

Sigma Xi. 1986. *Honor in Science*. New Haven, CT: Sigma Xi.

———. 1993. *Ethics, Values and the Promise of Science*. Research Triangle Park, NC: Sigma Xi.

Starr, P. 1982. *The Social Transformation of American Medicine*. New York: Basic Books.

Steelman, J. R. 1947. Science and public policy. The President's Scientific Research Board. Washington, DC: U.S. Government Printing Office.

Stokes, D. E. 1997. *Pasteur's Quadrant*. Washington, DC: Brookings Institution Press.

Swazey, J. P., M. S. Anderson, and K. S. Lewis. 1993. Ethical problems in academic research. *American Scientist* 81:542–553.

Thompson, D. F. 1993. Understanding financial conflicts of interest. *N. Engl. J. Med.* 329:573–6.

Whitbeck, C. 1996. Ethics as design: Doing justice to moral problems. *Hastings Center Report* 26:9.

Wright, R. 1994. *The Moral Animal*. New York: Vintage.

Ziman, J. M. 1968. *Public Knowledge*. Cambridge: Cambridge University Press.

———. 1978. *Reliable Knowledge*. Cambridge: Cambridge University Press.

Zuckerman, H. 1977. Deviant behavior and social control. In *Deviance and Social Control*, edited by E. Sagarin. Beverly Hills, CA: Sage Publications.

Zurer, P. S. 1987. Misconduct in research: It may be move widespread than chemists like to think *Chem. Eng. News* (13 April):10–17.

Index